William Edward Hartpole Lecky

The Leaders of Public Opinion in Ireland. Swift, Flood, Grattan, O'Connell

William Edward Hartpole Lecky

The Leaders of Public Opinion in Ireland. Swift, Flood, Grattan, O'Connell

ISBN/EAN: 9783337111809

Printed in Europe, USA, Canada, Australia, Japan

Cover: Foto ©ninafisch / pixelio.de

More available books at **www.hansebooks.com**

THE
LEADERS OF PUBLIC OPINION
IN
IRELAND:

SWIFT—FLOOD—GRATTAN—O'CONNELL.

BY

WILLIAM EDWARD HARTPOLE LECKY, M.A.

'The breath of Liberty, like the word of the holy man, will not die with the prophet, but will survive him.'—GRATTAN.

NEW EDITION,

REVISED AND ENLARGED.

LONDON:
LONGMANS, GREEN, AND CO.
1871

CONTENTS.

	PAGE
JONATHAN SWIFT	1
HENRY FLOOD	63
HENRY GRATTAN	104
DANIEL O'CONNELL	223

INTRODUCTION.

IN REPUBLISHING the following sketches, which first appeared anonymously many years ago, I am yielding in part to the request of many friends in Ireland and elsewhere who have been good enough to regret the difficulty of procuring them; and in part also to a feeling that at the present moment their appearance might not be wholly useless or inopportune. At a time when the Repeal movement which was suspended by the famine is manifestly reviving; when the establishment of religious equality has removed the old lines of party controversy, and prepared the way for new combinations; when security of tenure, increased material prosperity, the spread of education, and the approaching triumph of the ballot, have given a new weight and independence to the masses of the people; and when, at the same time, a disloyalty in some respects of a more malignant type than that of any former period has widely permeated their ranks, it is surely not unadvisable to recall the leading facts of the great struggle of Irish nationality. The present of a nation can only be explained by its past; and in dealing with strong sentiments of disloyalty and discontent, it is of the utmost importance to trace the historical causes to which they may be due.]

There are no errors in politics more common or more fatal than the political pedantry which estimates institutions exclusively by their abstract merits, without any regard to the special circumstances, wishes, or characters of the nations for which they are intended, and the political materialism which refuses to recognise any of what are called sentimental grievances. Political institutions are essentially organic things, and their success depends, not merely on their intrinsic excellence, but also on the degree in which they harmonise with the traditions and convictions, and take root in the affections of the people. Every statesman who is worthy of the name will carefully calculate the effect of his measures upon opinion, will esteem the creation of a strong, healthy, and loyal public spirit one of the highest objects of legislation, and will look upon the diseases of public opinion as among the greatest evils of the State.

[There is, perhaps, no government in the world which succeeds more admirably in the functions of eliciting, sustaining, and directing public opinion than that of England. It does not, it is true, escape its full share of hostile criticism, and, indeed, rather signally illustrates the saying of Bacon, that 'the best governments are always subject to be like the finest crystals, in which every icicle and grain is seen which in a fouler stone is never perceived;' but whatever charges may be brought against the balance of its powers, or against its legislative efficiency, few men will question its eminent success as an organ of public opinion.] In England an even disproportionate amount of the national talent takes the direction of politics. The pulse of an energetic national life is felt in every quarter of the land. [The debates of Parliament are followed with a warm, constant, and intelligent interest by all sections of the community. It draws all classes within the circle of political interests, and is the centre of a strong and steady patriotism equally removed from the apathy of many continental nations in time of calm, and from their feverish and spasmodic energy in time of excitement. Its decisions, if not instantly accepted, never fail to have a profound and a calming influence on the public mind. It is the safety-valve of the nation. The discontents, the suspicions, the peccant humours that agitate the people find there their vent, their resolution, and their end.]

It is impossible, I think, not to be struck by the contrast which in this respect Ireland presents to England. If the one country furnishes us with an admirable example of the action of a healthy public opinion, the other supplies us with the most unequivocal signs of its disease. The Imperial Parliament exercises for Ireland legislative functions, but it is almost powerless upon opinion—it allays no discontent, and attracts no affection. Political talent, which for many years was at least as abundant among Irishmen as in any equally numerous section of the people, has been steadily declining; and the marked decadence in this respect among the representatives of the nation reflects but too truly the absence of public spirit in their constituents. [The upper classes have lost their sympathy with and their moral ascendency over their tenants, and are thrown for the most part into a policy of mere obstruction. The genuine national enthusiasm never flows in the channel of imperial politics. With great multitudes sectarian considerations have entirely superseded national ones, and their representatives are accustomed systematically to subordinate all party and all political questions to ecclesiastical interests; and while calling themselves Liberals, they make it the main object of their home politics to separate the different classes of their fellow-countrymen during the period of their education, and the main object of their foreign policy to support the temporal power of the Pope. With another and a still larger class the prevailing feeling seems to be an indifference to all Parliamentary proceedings; an utter scepticism about constitutional means of realising their ends; a blind, persistent hatred of England. Every cause is taken up with an enthusiasm exactly proportioned to the degree in which it is supposed to be injurious to English interests. An amount of energy and enthusiasm which if rightly directed would suffice for the political regeneration of Ireland is wasted in the most insane projects of disloyalty; while the diversion of so much public feeling from Parliamentary politics leaves the Parliamentary arena more and more open to corruption, to place-hunting, and to imposture.]

This picture is in itself a very melancholy one, but there are other circumstances which greatly heighten the effect. In a very ignorant or a very wretched population it is natural that there should be much vague, unreasoning discontent; but the Irish people are at present neither wretched nor ignorant. Their economical condition before the famine was indeed such that it might well have made reasonable men despair. With the land divided into almost microscopic farms, with a population multiplying rapidly to the extreme limits of subsistence, accustomed to the very lowest standard of comfort, and marrying earlier than in any other northern country in Europe, it was idle to look for habits of independence or self-reliance, or for the culture which follows in the train of leisure and comfort. But all this has been changed. A fearful famine and the long-continued strain of emigration have reduced the nation from eight millions to less than five, and have effected, at the price of almost intolerable suffering, a complete economical revolution. The population is now in no degree in excess of the means of subsistence. The rise of wages and prices has diffused comfort through all classes. [The greater part of Ireland has been changing from arable into pasture land, for which it is pre-eminently fitted; and this most important transformation, which almost convulsed English society in the sixteenth century, and elicited the bitterest lamentations from Bacon and More, has been of late years effected in Ireland upon a still larger scale without producing any considerable suffering. It is following in the train of a natural movement of emigration, springing no longer from distress or from landlord tyranny, but partly from a healthy spirit of industrial ambition impelling young men to the great fields of enterprise in the new world with which they are no longer unacquainted, and partly from a feeling of natural affection drawing the older members of a family to the distant homes which their children have established.] Probably no country in Europe has advanced so rapidly as Ireland within the last ten years, and the tone of general cheerfulness, the improvement of the houses, the dress, and the general condition of the people must have struck every observer. Ireland is no doubt still very poor if compared with England, or even with Scotland; but its poverty consists much more in the absence of great wealth than in the presence of great misery. It has been recently stated that while paupers are in England as one to twenty, and in Scotland as one to twenty-three of the population, in Ireland they are only as one to seventy-four.[1] At the same time industrial habits have been rapidly spreading. The custom of early marriages, which lay at the root of the economical evils of Ireland, has, according to recent statistics, been seriously checked; and the standard of comfort is far higher and the spirit of industrial progress far more active than in any previous portion of the century. If industrial improvement, if the rapid increase of material comforts among the poor, could allay political discontent, Ireland should never have been so loyal as at present.

Nor can it be said that ignorance is at the root of the discontent. The Irish people have always, even in the darkest period of the penal laws, been greedy for knowledge, and few races show more quickness in acquiring it. The admirable system of national education established in the present century is beginning to bear abundant fruit, and among the younger generation at least, the level of knowledge is quite as high as in England. Indeed, one of the most alarming features of Irish disloyalty is its close and evident connection with education. It is sustained by a cheap literature, written often with no mean literary skill, which penetrates into every village, gives the people their first

[1] See 'Fawcett on Pauperism,' p. 27.

political impressions, forms and directs their enthusiasm, and seems likely in the long leisure of the pastoral life to exercise an increasing power. Close observers of the Irish character will hardly have failed to notice the great change which since the famine has passed over the amusements of the people. The old love of boisterous out-of-door sports has almost disappeared, and those who would have once sought their pleasures in the market or the fair now gather in groups in the public-house, where one of their number reads out a Fenian newspaper. Whatever else this change may portend, it is certainly of no good omen for the future loyalty of the people.

It was long customary in England to underrate this disaffection by ascribing it to very transitory causes. The quarter of a century that followed the Union was marked by almost perpetual disturbance, but this it was said was merely the natural ground-swell of agitation which followed a great reform. It was then the popular theory that it was the work of O'Connell, who was described during many years as the one obstacle to the peace of Ireland, and whose death was made the subject of no little congratulation, as though Irish discontent had perished with its organ. It was as if, the Æolian harp being shattered, men wrote an epitaph upon the wind. Experience has abundantly proved the folly of such theories. Measured by mere chronology, a little more than seventy years have passed since the Union; but famine and emigration have compressed into those years the work of centuries. The character, feelings, and conditions of the people have been profoundly altered. A long course of remedial legislation has been carried, and during many years the national party has been without a leader and without a stimulus. Yet, so far from subsiding, disloyalty in Ireland is probably as extensive, and is certainly as malignant, as at the death of O'Connell, and in many respects the public opinion of the country has palpably deteriorated. [O'Connell taught an attachment to the connection, a loyalty to the Crown, a respect for the rights of property, a consistency of Liberalism, which we look for in vain among his successors; and that faith in moral force and constitutional agitation which he made it one of his greatest objects to instil into the people has almost vanished with the failure of his agitation.]

[The causes of this deep-seated disaffection I have endeavoured in some degree to investigate in the following essays.] To the merely dramatic historian the history of Ireland will probably appear less attractive than that of most other countries, for it is somewhat deficient in great characters and in splendid episodes; but to a philosophic student of history it presents an interest of the very highest order. [In no other history can we trace more clearly the chain of causes and effects, the influence of past legislation, not only upon the material condition, but also upon the character of a nation. In no other history especially can we investigate more fully the evil consequences which must ensue from disregarding that sentiment of nationality which, whether it be wise or foolish, whether it be desirable or the reverse, is at least one of the strongest and most enduring of human passions. This, as I conceive, lies at the root of Irish discontent. It is a question of nationality as truly as in Hungary or in Poland. Special grievances or anomalies may aggravate, but do not cause it, and they become formidable only in as far as they are connected with it. What discontent was felt against the Protestant Established Church was felt chiefly because it was regarded as an English garrison sustaining an anti-national system; and the agrarian difficulty never assumed its full intensity till by the Repeal agitation the landlords had been politically alienated from the people.

The evils of the existing disloyalty are profoundly felt in both nations. Nature and a long and inextricable union of interests have made it imperatively necessary for the two countries to continue under the same rule. No reasonable man who considers their relative positions can believe that England would ever voluntarily relinquish the government of Ireland, or that Ireland could ever establish her independence in opposition to England, unless the English navy were utterly shattered. Even in the event of the dissolution of the Empire, Irish separation could only be achieved at the expense of a civil war, which would probably result in the massacre of a vast section of the Irish people, would drive from the country much of its intelligence and most of its capital, and would inevitably and immediately reduce it to a condition of the most abject misery. Nor would any class suffer more than the class by which revolutions are usually made. For poor men of energy and talent, the magnificent field of Indian and colonial administration, which is now thrown open to competition, offers a career of ambition incomparably surpassing in splendour that which any other European nation can furnish; and Irishmen have fully availed themselves of it. Though their country is but a small one, it now plays no inconsiderable part among men; for while Irish emigration is leavening the New World, Irish administrators under the British Crown are organising in no small degree the empires or the republics of the future. All this noble career for talent and enterprise would be destroyed by separation; every element of Irish greatness would dwindle or perish; the energies of the people, confined to the narrow circle of a small and isolated State, would be wasted in petty quarrels, sink into inanity, or degenerate into anarchical passions.

These would be the consequences of the separation of Ireland from the British Empire. That such a severance is almost impossible may be readily admitted; but still, in a great European convulsion, Ireland might be a serious danger to England. [Even in time of peace its discontent necessitates a heavy military expenditure, and the emigration from its shores is multiplying enemies to England through the New World. In foreign policy it is a manifest source both of weakness and discredit. For many years English Liberals have made it a main principle of their foreign policy to advocate the settlement of all disputes between rulers and their subjects in accordance with the desires of the latter; and the fact that in a portion of their own country the existing form of government is notoriously opposed to the wishes of the people supplies their adversaries with an obvious answer. In home politics, the presence in Parliament of a certain number of members who are alienated from the general interests of the Empire, and actuated by a spirit out of harmony with that of the constitution, is a serious danger; and it acquires additional gravity as parties disintegrate or tend to equilibrium. It lowers the tone of Liberalism, leads to unnatural coalitions and surprises, and is a constant temptation to rival leaders to purchase this support by unworthy concessions. Apart from the possible horrors of rebellion, the mere existence of a widespread disloyalty restricts the flow of capital which is essential to the full development of Irish resources; and the direction of so large an amount of the enthusiasm of the country in opposition to the law, and the diversion of much more into sectarian channels, vitiates and debases all political life. At the same time a constant fever of political agitation is sustained.] For a long time it was the custom to send to Ireland officials who were utterly inexperienced, or who, on account of their characters, would have been tolerated nowhere else. This system, which O'Connell compared to that of country barbers making their apprentices take their first lessons in shaving upon a beggar, and which in the last century elicited a very striking protest from Lord Northington,[1] can hardly be said to continue, but an equally mischievous one remains. The Irish difficulty has an irresistible attraction to party leaders who desire to raise some question that may embarrass or displace a Ministry—to theorists who have crotchets to display or political experiments to try—to revolutionists who wish to set in motion some subversive policy which they think may eventually be extended to

---

[1] When appointed Lord-Lieutenant in 1783, he wrote to Fox: 'I must confess that it is a very wrong measure of English government to make this country their first step in politics, as it usually has been; and I am sure men of abilities, knowledge, business, and experience ought to be employed here, both in the capacity of Lord-Lieutenant and Secretary, not gentlemen taken wild from Brookes's to make their début in public life.'—*Lord Russell's Life of Fox*, vol. ii. p. 23.

England. Writers who have never even crossed the Channel, and who are totally unacquainted with the practical working of Irish institutions and with the character of the people, dogmatise on the subject with the utmost confidence, and throw in fresh brands of discord at every period of crisis. More perhaps than anything else, the country needs repose, but, in addition to its own elements of anarchy, a torrent of irritating extraneous influences is constantly agitating it.

The three great requisites of good government for Ireland are that it should be strong, that it should be just, and that it should be national. It should be strong as opposed to that miserable system which resists every measure of popular demand as long as the country is quiet, and then concedes it without qualification as the prize of disloyalty and crime, and which has made it a settled maxim among Irishmen that the favours of the Government are bestowed upon every class in direct proportion to the dangers that are apprehended from it. It should be just as opposed to that system which at one time leans wholly to Catholics or to tenants, and at another time wholly to Protestants or to landlords, which will suffer an illegal procession in one province that would be rigidly repressed in another, and which subordinates all questions of patronage or principle, and even in some instances the very execution of the laws, to the exigencies of party politics. By such systems the respect for law has been fatally weakened, and their abandonment is the first condition of political health. But, in addition to this, it appears to me to be perfectly evident, from the existing state of public opinion in Ireland, that no Government will ever command the real affection and loyalty of the people which is not in some degree national, administered in a great measure by Irishmen and through Irish institutions. If the present discontent is ever to be checked, if the ruling power is ever to carry with it the moral weight which is essential to its success, it can only be by calling into being a strong local political feeling, directed by men who have the responsibility of property, who are attached to the connection, and who at the same time possess the confidence of the Irish people. As in Hungary, as in Poland, as in Belgium, national institutions alone will obtain the confidence of the nation, and any system of policy which fails to recognise this craving of the national sentiment will fail also to strike a chord of true gratitude. It may palliate, but it cannot cure. It may deal with local symptoms, but it cannot remove the chronic disease. To call into active political life the upper class of Irishmen, and to enlarge the sphere of their political power—to give, in a word, to Ireland the greatest amount of self-government that is compatible with the unity and the security of the Empire—should be the aim of every statesman.

To do this is, unfortunately, extremely difficult. At present the very materials and essential conditions of self-government are in a great degree wanting. There was a time when the attachment of the occupiers of the soil to their landlords was probably as warm in Ireland as in any other country, but a long series of causes, which I have endeavoured to trace in the following pages, have greatly diminished it, and the schism of classes, and the wild notions on the subject of landed property which have of late years been diffused, constitute a serious danger. The motives of interest that connect Ireland with England are sufficient to secure the co-operation of the two countries as long as Irish opinion is directed by property and intelligence, but they are not likely to weigh with unprincipled adventurers, or with ignorant and unreasoning disloyalty. [At the same time, sectarian feeling runs so high in politics that it is probable that one of the first acts of an Irish Parliament would now be to build up a wall of separation between Protestants and Catholics by the destruction of united education. Under such circumstances a sudden change of system is probably to be deprecated, and it is only by slow, cautious, and gradual steps that self-government can be in some degree restored. By steadily opposing the tendency to centralisation, which has produced so many evils in Ireland, by transferring private business from the overworked Parliament of the Empire to cheaper and perhaps more competent local tribunals, by gradually enlarging the sphere of local government, and by encouraging and bringing into activity the political talent of the country, a sound public opinion may be slowly formed.] Local government in Ireland, in as far as it exists, presents on the whole a very remarkable and very satisfactory contrast to the political condition of the country. The public institutions are probably quite as well managed as those of England, or indeed of any other country. The magistracy, the police, and the poor-law administration are eminently efficient, and the comparatively small amount of pauperism is partly due to the good management of the latter. One of the latest signs of the deplorable local government in England has been the epidemic of small-pox which has followed the general neglect of the law about vaccination; but in Ireland no such epidemic has raged, and the fact is ascribed chiefly to the much better enforcement of the law. One of the most important recent movements in the direction of prison reform has been due to the success of the reformatory system which has been established in Ireland. Undetected agrarian crime, the untrustworthiness of juries in cases on which public feeling is strongly excited, the scandalous tone of a certain section of the press, and the frequency of religious or political riots still disgrace the country; but the first and last of these evils have been restricted within very narrow territorial limits; the second might be greatly mitigated by the introduction of the Scotch jury system, under which unanimity is not necessary for a verdict; and the general average both of serious crime and of vice is lower than in England. It would be a gross injustice to the country to infer that its political condition reflects accurately its social condition, or that the relations of landlords and tenants are habitually hostile. If the people are deficient in self-reliance, they are at least eminently susceptible to discipline, their natural instincts are aristocratic, and they are very faithful to their leaders.

If it be true that the desire for some measure of self-government is not likely to be extinguished or diminished in Ireland, it is evident that many influences are in operation which must tend towards its realisation. Of the two great Irish measures which have been passed within the last few years, it will probably be found that the one disestablishing the Protestant Church will have effects little contemplated by the bulk of its supporters. The question was always mainly an English one. Since the tithes were commuted into a land-tax, paid exclusively by the landlords, the great body of the Irish people have cared very little on the subject. The Protestant clergy were usually popular and useful; with the exception of priests and converts, few people in Ireland grudged them their endowments; and if it had not been for English party interests, and for the radicalism of British Dissent, they might long have continued. If, indeed, the Church funds had been divided between the rival sects, the conciliatory effect of the measure might have been very great. The partial payment of the priests—which a long series of eminent statesmen of different parties, from Pitt to Lord Russell, have concurred in recommending—would have attached the most influential class in Ireland indissolubly to the throne, would have appreciably raised their social position, and, by relieving the poorer Catholics of their most oppressive burdens, would have been felt with gratitude in every household. If the independence of the priesthood had been fully guaranteed, the Irish objections to such a measure would probably have been surmounted; but English, and especially Scotch, public opinion made it impossible. The Radicals, who desired the abolition of the Irish Establishment mainly as a step to the abolition of the English one—the Puritans, whose hatred of Catholicism was even stronger than their hatred of Establishments—interposed their veto, and the Church Bill was carried in a form which was of little or no practical benefit to the Catholics, who have accordingly received it with general indifference. But its effect upon the Protestants has been extremely great. They have been cut loose from their old moorings. The object the defence of which was a main end of their policy has disappeared, and they are certainly more disposed than at any period since the Union to throw themselves into the general current of Irish sentiment. At the same time, the representative bodies in which the Irish gentry are learning to assemble to deliberate upon their Church affairs are forming habits which may be extended to politics. In spite of frequent and menacing reactions, it is probable that sectarian animosity will diminish in Ireland. The general intellectual tendencies of the age are certainly hostile to it. With the increase of wealth and knowledge there must in time grow up among the Catholics an independent lay public opinion, and the tendency of their politics will cease to be purely sacerdotal. The establishment of perfect religious equality and the settlement of the question of the temporal power of the Pope have removed grave causes of irritation, and united education, if it be steadily maintained and honestly carried out, will at length assuage the bitterness of sects and perhaps secure for Ireland the inestimable benefit of real union. The division of classes is at present perhaps a graver danger than the division of sects. But the Land Bill of Mr. Gladstone cannot fail in time to do much to cure it. If it be possible in a society like our own to create a yeoman class intervening between landlords and tenants, the facilities now given to tenants to purchase their tenancies will create it; and if, as is probable, it is economically impossible that such a class should now exist to any considerable extent, the tenant class have at least been given an unexampled security—they have been rooted to the soil, and their interests have been more than ever identified with those of their landlords. The division between rich and poor is also rapidly ceasing to coincide with that between Protestant and Catholic, and thus the old lines of demarcation are being gradually effaced. A considerable time must elapse before the full effect of these changes is felt, but sooner or later they must exercise a profound influence on opinion; and if they do not extinguish the desire of the people for national institutions, they will greatly increase the probability of their obtaining them.

L.

THE
LEADERS OF PUBLIC OPINION
IN
IRELAND.

JONATHAN SWIFT.

JONATHAN SWIFT was born in Dublin in the year 1667. His father (who had died a few months before) had been steward of the King's Inn Society. His mother was an English lady of a Leicestershire family, remarkable for the strictness of her religious views, and for the energy and activity of her character. At the early age of six, Swift was sent to a school at Kilkenny, where he remained till he was fourteen, when he entered the University of Dublin. His position there was exceedingly painful, and he remembered it with bitterness to the end of his life. His sole means of subsistence were the remittances of his uncle Godwin: and those remittances, owing to the poverty—or, as Swift believed, the miserly disposition—of his uncle, were doled out in the most niggardly manner. He found it impossible to maintain the position of a gentleman. He was precluded from all the luxuries, and could with difficulty procure the necessaries of life.

Notwithstanding the extreme frugality with which he managed his slender resources, he was on one occasion left absolutely destitute, and was relieved only by the unexpected arrival of a present from a cousin, who was a merchant at Lisbon. The conduct of a young man under such circumstances often furnishes no obscure intimation of the prevailing character of his after-life. Goldsmith, when struggling with extreme poverty, at the University, lived in the most reckless enjoyment, spending what money he had with profuse generosity, disregarding as far as possible the studies of his course, and only employing his fine talents in writing street-ballads, which he sold to supply his more pressing wants. Johnson, in a similar position, grew morose, and turbulent, and domineering. He defied the discipline, but availed himself fully of the intellectual advantages of, college, and astonished and delighted his tutors by the extent and the accuracy of his information.

Swift, like Johnson, was completely soured by adversity, and, like Goldsmith, he treated the academic studies with supreme contempt. He systematically violated all college rules—absenting himself from night-roll, chapel, and lectures, haunting public-houses, and in every way defying discipline. He considered mathematics, logic, and metaphysics useless, and accordingly positively refused to study them. Dr. Sheridan (who was a good mathematician) tells us that in after-life he had attained some proficiency in the first of these subjects, but the hatred and contempt he entertained for it never diminished. His ignorance of logic was so great that at his degree examination he could not even frame a syllogism, and accordingly was unable to pass the examination, and only obtained his degree 'by special favour'—a fact which is still remembered with pleasure by the undergraduates who are examined beneath his portrait. Yet, even at this time, his genius was not undeveloped or unemployed. He studied history, he wrote odes, and, above all, he composed his 'Tale of a Tub.' The first draft of this wonderful book he showed to his college friend Warren when he was only nineteen, but he afterwards amplified and revised it considerably, and its publication did not take place till 1704. He also acquired at this time those pedestrian habits which continued through life, and exercised so great an influence upon his mind. He traversed on foot a considerable portion of England and Ireland, mingling with the very lowest classes, and sleeping at the lowest public-houses. The traces of this habit may be seen on almost every page of his writings. To this period of his life we probably owe the taste for coarse, vulgar illustrations, by which his noblest works are disfigured, as well as much of that minute observation, that keen and accurate knowledge of men, which is one of their greatest charms. To the end of his life he delighted in mixing with men of the lowest classes, and no great writer ever understood better the art of adapting his style to their tastes and understandings. To the same period of his life we may trace the careful and penurious habits which in his old age developed into an intense avarice.

Upon leaving the University, the first gleam of prosperity, though at first hardly of happiness, shone upon his path. His mother was related to the wife of Sir W. Temple, and this circumstance procured for him the position of amanuensis at Moor Park, which he held for several years.

Sir W. Temple was at this time near the close of his career. He enjoyed the reputation of a considerable statesman and of a very great diplomatist, and his character was in truth much more suited for negotiation than for the rougher forms of statesmanship. [With great abilities and much kindness of heart, he was too languid, unambitious, and epicurean to attain the highest place in English politics; and his bland, patronising courtesy, his refined and somewhat fastidious taste, as well as his instinctive shrinking from turmoil, controversy, and violence, denoted a man who was more fitted to shine in a court than in a parliament. He described in one of his Essays 'coolness of temper and blood, and consequently of desires,' as 'the great principle of virtue,' and his disposition almost realised his ideal. He had, however, a considerable knowledge of men and books, and a sound and moderate judgment in politics; and his life, if it was distinguished by no splendid virtues, and characterised by a little selfishness and a little cowardice, was at least singularly pure in an age when political purity was very rare.] He had surrounded himself in his old age with beautiful gardens, and objects of art and refinement; and he dallied in a feeble way with literature, writing in admirably pure, graceful, and melodious English, somewhat vapid essays on politics and gardens, on Chinese literature and the Evil of Extremes.

With a character of this kind Swift could have little sympathy. For good or for evil, intensity was always one of his leading characteristics. It was shown alike in his friendships and his enmities, in his ambitions and his regrets. Though not susceptible to the common passion of love, a liquid fire seemed coursing through his veins. That 'sæva indignatio' which he recorded in his epitaph, the fierce ambition, the indomitable pride, the intense hatred of wrong, which he invariably displayed, must have often made him strangely at variance with his courtly patron. His position was extremely galling, for he was at first only treated as a kind of upper servant. He was shy and awkward, and felt, as he afterwards confessed, keenly a word of disapprobation from Temple. His college habits doubtless gave an additional roughness to his manners; and the ill health, which had already begun to prey upon him, an additional acerbity to his temper.

However, as time advanced, his position at Moor Park improved. He devoted himself most assiduously to study for several years, and thus compensated for his idleness at the University. His favourite subjects appear to have been the classics and French literature; and he read them with the energy of enthusiasm. In 1692 he took his degree of Master of Arts at Oxford, for which University he ever after entertained feelings of grateful regard. He also rose rapidly in Sir W. Temple's estimation, and hoped, through his influence, soon to obtain an independent position. He believed, however (whether justly or unjustly we need not too curiously enquire), that Temple's patronage was very languid, and he at last left Moor Park in anger, and proceeded to Ireland to be ordained. He there found, to his inexpressible dismay, that a letter of recommendation from Temple was an indispensable preliminary to ordination. For months he shrank from the humiliation of asking the letter, but at last he wrote for and received it. He was ordained, and almost immediately after he obtained a small preferment at a place called Kilroot, in the diocese of Connor. Temple, however, in the meantime, had found that Swift's presence was absolutely necessary to his enjoyment. The extreme amiability of his disposition prevented him from retaining any feelings of bitterness, and he made overtures which soon drew the young clergyman from a retirement that was as unsuited to his happiness as to his genius. Swift returned to England, and lived with Temple till the death of the latter, which took place four years after. During this time he was treated not as a dependent, but as a friend. He was admitted into his patron's confidence; his genius was fully recognised; and the bias of his mind determined for life. Living with an old statesman of great experience, sagacious judgment, and varied knowledge, it was natural that his attention should be chiefly turned to politics. His first pamphlet—the 'Dissentions of the Nobles and Commons of Athens'—was published somewhat later in the Whig interest. It was extremely successful, and was generally attributed to Bishop Burnet. He had several opportunities of seeing the King, and some of the leading statesmen of the day, who visited Moor Park—of gauging their intellects, and correcting his theories by their experience.

On one occasion he was deputed by Temple to endeavour to persuade the King to consent to triennial parliaments—a mission in which he did not succeed. He also attended largely to literature. He assisted Temple in revising his works, and he defended him against the well-known assaults of Bentley. Temple had rashly committed himself to the authenticity of some spurious letters attributed to Phalaris, and had launched into a eulogium of these letters in particular, and generally of ancient as opposed to modern literature. The dispute had been warmly taken up by Boyle and Atterbury on one side, and by Bentley on the other. The scholarship of Bentley proved overwhelming, and his opponents were at last driven from the field; but Swift, avoiding judiciously all direct argumentative collision with so formidable an opponent, produced his 'Battle of the Books,' which was then and is now unrivalled in its kind. But it was not merely the gratification of political or literary ambition that made the last portion of Swift's residence at Moor Park so attractive. That strange romance which tinged all his later years had begun, and his life was already indissolubly connected with that of Esther Johnson.

Esther Johnson, so well known by the name of Stella, was the reputed daughter of the steward of Sir W. Temple, but many persons maintained that Temple himself was her father, and they imagined they could detect the parentage in her features. The peculiar position she seems to have occupied at Moor Park, and the large legacy left her by Temple, go far to corroborate the supposition. At the time we speak of she was in the very zenith of her charms. Her figure, which in after-years lost much of its grace and symmetry, was then faultless in its proportions, and her biographers dilate with rapture on the intellectual beauty of her pale but not pensive countenance, shadowed by magnificent raven hair, and illumined by dark, lustrous, and trembling eyes. Her temperament was singularly serene, patient, and unimpassioned, admirably suited for social life, and for sustained friendship, but a little too cold for real love, and she appears to have acquiesced for many years, without repining, in a kind of connection which few women would have tolerated. But great as were her personal charms, her intellectual gifts were far more remarkable, and she seems to have lived more from the head than from the heart. She had read much and in many fields, and her wit made her the delight of every society in which she moved. Swift said that in whatever company she appeared it seemed to be invariably admitted that she had said the best thing of the evening, and though the witticisms he has preserved exhibit quite as much coarseness as point, her principal extant poem—that to Swift on his birthday in 1721—fully sustains her reputation.[1]

[I do not intend in the present sketch to enter at length into an examination of the controversy about the nature of the connection that subsisted for so many years between Swift and Esther Johnson. Such matters are perhaps given a rather disproportionate place in the lives of men of genius; and, at all events, the object of this work is to deal with the political aspects of his career. There appears, however, to be no real doubt that that connection was always purely platonic. They lived in Ireland in different houses, except during the illnesses of Swift. Stella presided at the table of Swift when he received company. Their correspondence was of the most affectionate character, and Stella has acquired an immortality of fame through the poetry of her friend. At the same time, that poetry, though indicating the affection of a warm friend, is wholly unlike that of a lover, and it is curious to observe how constantly Swift decries her personal beauty, and directs his most graceful compliments to her other qualities.]

> But, Stella, say what evil tongue
> Reports that you're no longer young;
> That Time sits with his scythe to mow
> Where erst sat Cupid with his bow;
> That half your locks are turned to grey.
> I'll ne'er believe a word they say!
> 'Tis true—but let it not be known—
> My eyes are somewhat dimmish grown;
> For Nature, always in the right,
> To your defects adapts my sight;

---

[1] There is one other short poem, 'Lines to Jealousy,' ascribed to her.

> And wrinkles undistinguished pass,
> For I'm ashamed to use a glass;
> And till I see them with these eyes,
> Whoever says you have them, lies.
> No length of time can make you quit
> Honour and virtue, sense and wit:
> Thus you may still be young to me,
> While I can better hear than see.
> Oh ne'er may Fortune show her spite
> To make me deaf and mend my sight!

Upon the death of Temple, Swift was once more thrown upon the world, but his prospects were exceedingly favourable. Temple (who during his long, painful illness, had found Swift unwearied in his attention) had taken every means of ensuring his future prosperity. He left him a pecuniary legacy, together with the charge and profit of publishing his posthumous works, and he had procured for him from King William a promise of a prebend either at Canterbury or Windsor.

Temple's posthumous works were rapidly published and dedicated to the King, who, however, took no notice of the dedication, of his old servant's request, or of his own promise. Shortly afterwards, Swift obtained the position of secretary to the Earl of Berkeley, who had been appointed one of the Lords Justices in Ireland; but a person named Bushe succeeded in persuading the Earl that the office should not be held by a clergyman, and in obtaining it for himself. Another disappointment followed. He was almost appointed to the deanery of Down, but the appointment was stayed by the interposition of Archbishop King, who objected to his extreme youth. Lord Berkeley, as if to compensate for these disappointments, then gave him the living of Laracor and Rathbeggan. He remained for some time at Laracor in the discharge of his clerical duties; and Stella, accompanied by a Mrs. Dingle, a lady of a very negative character, came over and resided near him. Before long, however, he was called from his country living to partake in the great political struggles of the day.

In 1710 the Primate of Ireland sent him to London' to endeavour to procure a remission of the payment to the Crown by the Irish clergy of the first fruits and the twentieth parts. He succeeded in his mission, and he, at the same time, found himself drawn into the vortex of politics.

The Whig ministry, under Somers and Godolphin, had just fallen. Harley and St. John, the leaders of the Tories, had succeeded them, but their position was exceedingly precarious. The feelings of the people were against them. The chief political writers of the day assailed them with unsparing severity; and the Queen had, on at least one occasion, slighted them in the most undisguised manner. The age, as Macaulay observes, was essentially an age of essays. The press was yet undeveloped, the speeches of Parliament were unreported, but yet a strong intellectual energy pervaded the nation. Under these circumstances the writers of pamphlets, or of short political essays, like the 'Examiner,' were the real rulers of England. In the composition of these essays Swift was unrivalled, except by Addison, and scarcely equalled by him.

The Whigs naturally supposed that he would devote his talents to their service, but they soon found that they were mistaken. Swift treated them with marked coldness. He refused, at Lord Halifax's, to drink the 'resurrection' of the Whigs, unless it were accompanied by their reformation; and he at length openly joined himself to the Tories. The reasons he assigned for this change were very simple. He had originally been a Whig because he justified the Revolution, which could only be defended on Whig principles. On the other hand, as a clergyman and a High Churchman, he considered the exclusion of Dissenters from State offices essential to the security of the Church, and he therefore abandoned the Whigs, who had constituted themselves the champions and representatives of the Dissenting interest. At the same time he more than once avowed, with that curious frankness for which he was remarkable, that personal motives contributed to his change. Godolphin had treated him with great coldness; he had been neglected and disappointed by the party; he considered that no personal obligation bound him to the falling fortunes of the Whigs; and he met with warm encouragement from Harley and St. John, the leaders of the Tories. He was very poor, very able, and very ambitious, and his interests and his sympathies tended in the same direction.

This change, as might have been expected, has exposed Swift to bitter attacks from most Whig and from some Tory writers—attacks that have been the more natural because Tory principles have found no abler defender, and Whig statesmen no more rancorous assailant, than this former Whig. But although in this as in most periods of his life Swift acted through mixed motives, I do not think that an impartial judge will pronounce any very severe sentence upon it. It was almost inevitable that a young man brought up in the house of Sir W. Temple should begin his career as a Whig. It was almost equally certain that a High Church clergyman would ultimately gravitate to the Tories. Swift, though he disliked William, never appears to have questioned the necessity of the Revolution, and in this respect he continued a Whig. Nor was he ever implicated, like his

Tory friends, in negotiations with the Pretender. But in the reign of Anne, and especially after the prosecution of Sacheverell had shattered the ministry of Godolphin, the great question dividing the two parties was not the question of dynasty, but the question of tests. It was much more a contest between the Church and Dissent than between the adherents of rival claimants to the throne. The ambiguous position and divided feelings of the Queen had suspended the conflict of the Revolution, and the injudicious prosecution of Sacheverell had aroused a spirit which entirely altered the relative positions of parties. The whole body of the Dissenters, and all who desired the repeal of the tests, supported the Whigs. The great majority of the Anglican clergy, and all the classes that were moved by the cry of 'Church in danger,' rallied round the Tories. It may appear strange that an intellect at once so powerful and so irreverent as that of Swift should have been wedded to High Church notions, but the fact is undoubted, and it is an entire misrepresentation to describe these sentiments as lightly or hastily assumed. The 'Tale of a Tub,' which was sketched in college, and published in 1704, shows all the Church principles and all the antipathy to Dissenters which he subsequently evinced. The same High Church principles appeared in a poem, which he wrote when with Sir W. Temple, in praise of Sancroft, in which he deplored the condition of the Church, 'led blindfold by the State.' In 1708 he published his 'Sentiments of a Church of England Man,' in which he describes himself as wavering between the parties, and aiming at neutrality, on the one hand justifying the Revolution, on the other deploring the prevailing sentiments about the Church. In a letter on the sacramental test, which appeared a few months later, he took a still stronger part against the Dissenters, and to this letter he ascribes the first coolness of his Whig friends. He said on one occasion that he could not understand a clergyman not being a High Churchman; and in every stage of his career he wrote steadily, persistently, and powerfully in favour of tests. In changing his side in politics he deserted men who had neglected and ill-treated him, but it would be difficult to show that he abandoned a single principle of secular politics, while he undoubtedly took the line in Church politics which his earliest writings had foreshadowed. No one, indeed, can compare his feeble essay on 'The Dissentions of the Nobles and Commons in Athens,' which is his one Whig pamphlet, with his later writings in defence of the Tories, without perceiving in which direction his mind naturally inclined. No doubt his junction with the Tories in 1710 was eminently to his advantage, but it should not be forgotten that in his later years he defended tests and disqualifications quite as jealously in Ireland at the very time when he was endeavouring to unite all Irishmen in their national cause. Such a bigotry is far from admirable, but it may at least claim the merit of sincerity.

The principal writers at this time on the Whig side were Addison, Steele, Burnet, Congreve, and Rowe, who were opposed by Atterbury, St. John, and Prior. Addison retired from the arena a few weeks before Swift entered it, and the latter was left without a rival. In many of the qualities of effective political writing he has never been surpassed. Without the grace and delicacy of Addison, without the rich imaginative eloquence or the profound philosophic insight of Burke, he was a far greater master of that terse, homely, and nervous logic which appeals most powerfully to the English mind, and no writer has ever excelled him in the vivid force of his illustrations, in trenchant, original, and inventive wit, or in concentrated malignity of invective or satire. With all the intellectual and most of the moral qualities of the most terrible of partisans he combined many of the gifts of a consummate statesman—a marvellous power of captivating those with whom he came in contact, great skill in reading characters and managing men, a rapid, decisive judgment in emergencies; an eminently practical mind, seizing with a happy tact the common-sense view of every question he treated, and almost absolutely free from the usual defects of mere literary politicians. But for his profession he might have risen to the very highest posts of English statesmanship, and in spite of his profession, and without any of the advantages of rank or office, he was for some time one of the most influential men in England. He stemmed the tide of political literature, which had been flowing strongly against his party, and the admirable force of his popular reasoning, as well as the fierce virulence of his attacks, placed him at once in the first position in the fray. The Tory party, assailed by almost overwhelming combinations from without, and distracted by the most serious divisions within, was sustained and defended by him. Its leaders were divided by interest, by temperament, and, in some degree, even by policy; but Swift's genius gained an ascendency over their minds, and his persuasions long averted the impending collision. Its extreme members had formed themselves into a separate body, and were clamouring for the expulsion of all Whigs from office; but Swift's Letter of Advice to the 'October Club' effected the dissolution of that body, and the threatened schism was prevented. The nation, dazzled by the genius of Marlborough, and fired by the enthusiasm of a protracted war, was fiercely opposed to a party whose policy was peace, but Swift's 'Examiners' gradually modified this opposition, and his 'Conduct of the Allies' for a time completely quelled it. The success of this pamphlet has scarcely a parallel in history. It seems to have for a time almost reversed the current of public opinion, and to have enabled the Ministers to conclude the Peace of Utrecht. Notwithstanding his coarseness and capricious violence, and an occasional eccentricity of manner which indicated not obscurely the seeds of insanity, the brilliancy of his matchless conversation made him the delight of every society, and his sayings became the proverbs of every coffee-house. Among his friends were men of all parties, of all creeds, and of all characters. In the course of a few years he was on most intimate terms with Addison and Steele, with Halifax, Congreve, Prior, Pope, Arbuthnot, and Peterborough, with Harley and St. John, and most of the other leaders of the day. In spite of the gloomy misanthropy of his temperament, and the savage recklessness with which he too often employed his powers of sarcasm, he was capable of splendid generosity, and of the truest and most constant friendship. Few men have ever obtained a deeper or more lasting affection, and we may well place the testimony of the illustrious men who knew him best in opposition to the literary judgments of posterity. 'Dear Friend,' wrote Arbuthnot in after-years, 'the last sentence of your letter plunged a dagger in my heart. Never repeat those sad but tender words, that you will try to forget me. For my part, I can never forget you—at least till I discover, which is impossible, another friend whose conversation could procure me the pleasure I have found in yours.' Addison termed him 'the most agreeable companion, the truest friend, the greatest genius of his age.' Pope after a friendship of twenty-three years wrote of him to Lord Orrery, 'My sincere love of that valuable, indeed incomparable man, will accompany him through life, and pursue his memory were I to live a hundred lives, as many of his works will live, which are absolutely original, unequalled, unexampled. His humanity, his charity, his condescension, his candour, are equal to his wit, and require as good and true a taste to be equally valued.'

Undoubtedly, in the first instance, many of these friendships arose from gratitude. Literature had not yet arrived at the period when it could dispense with patrons, and one of the legitimate goals to which every literary man aspired was a place under the State. This naturally drew the chief writers around Swift, and the manner in which he at this time employed his influence is one of the most pleasing features of his career. There is scarcely a man of genius of the age who was not indebted to him. Even his political opponents, even men who had written violently against his party, obtained places by his influence. Berkeley was drawn by him from the retirement of college, recommended more than once to the leading Tories, and placed upon the highway of promotion. Congreve was secured at his request in the place which the Whigs had given him. Parnell, Steele, Gay, Rowe, Phillips, and Diaper received places or other favours by his solicitation. He said himself, with a justifiable pride, that he had provided for more than fifty people, not one of whom was a relation. His influence in society as well as with the Government was ceaselessly employed in favour of literature. He founded the 'Scriblerus Club,' in which many of the chief writers of the day joined; he exerted himself most earnestly in bringing Pope forward, and obtaining subscriptions for his translation of Homer. He pressed upon the attention of the Government a plan (which is now, however, admitted to have been an unwise one) for watching over the purity of the language, and he on every occasion insisted on marked deference being paid to literary men. He himself took an exceedingly high, and indeed arrogant, tone with Harley and St. John; and when the former sent him a sum of money as a compensation for his services, he was so offended that their friendship was well-nigh broken for ever. That this tone was not, as has sometimes been alleged, the vulgar insolence of an upstart, is sufficiently proved by the deep attachment manifested towards him by both Harley and St. John long after their political connection had terminated.

During all this time Swift kept up a continual correspondence with Stella, in the shape of a Journal, recording with the utmost minuteness the events of every day. We have the clearest possible evidence that this Journal was not intended for any other eyes than those of Stella and Mrs. Dingle. It is filled with terms of the most childish endearment, with execrable puns, with passages written with his eyes shut, with extempore verses, and extempore proverbs; with the records of every passing caprice, of every hope, fear, and petty annoyance; and is evidently a complete transcript of his mind. In that Journal we can trace clearly the eminence to which he rose, and also the shadows that overcast his mind. One of the principal of these was the gradual decline of his friendship with Addison. Addison's habitual coldness had, at first, completely yielded to the charms of Swift's conversation, and notwithstanding the great dissimilarity of their characters, they lived on the most intimate terms. But Swift was a strong Tory, and Addison was a strong Whig; and Addison was almost identified with Steele, who was still more violent in his politics, and who, though he had received favours from Swift, had made a violent personal attack upon his benefactor,[1] and had elicited an equally violent reply: and these things tended to the dissolution of the friendship. There was never an open breach, but their intercourse lost its old cordiality, and the glow of affection that had once characterised it passed away never to return. 'I went to Mr. Addison's,' wrote Swift in his Journal, 'and dined with him at his lodgings. I had not seen him these three weeks; we are grown common acquaintance, yet what have I not done for his friend Steele! Mr. Harley reproached me the last time I saw him, that, to please me, he would be reconciled to Steele, and had promised and appointed to see him, and that Steele never came. Harrison, whom Mr. Addison recommended to me, I have introduced to the Secretary of State, who has promised me to take care of him; and I have represented Addison himself so to the Ministry, that they think and talk in his favour, though they hated him before. Well, he is now in my debt—there is an end; and I never had the least obligation to him—and there is another end.'

Another source of annoyance to Swift was the difficulty with which he obtained Church preferment. He knew that his political position was necessarily exceedingly transient; he had no resources except his living, and he was extremely ambitious. By his influence at least one bishopric and innumerable other places had been given away, and yet he was unable to obtain for himself any preferment that would place him above the vicissitudes of politics. The reason of this was, that the Queen had conceived an intense antipathy to him.

[1] In a pamphlet called 'The Crisis.'

Sharpe, the Archbishop of York, had shown her his 'Tale of a Tub,' and had represented him as an absolute free thinker; the Duchess of Somerset, whose influence at court was very great, and whom he had bitterly and coarsely satirised, employed herself with untiring hatred in opposing his promotion; and the impression they made on the mind of Anne was such that all the remonstrances of the Ministers and all the entreaties of Lady Masham were unable to overcome it.

The charge of scepticism has been frequently reiterated in the present day, and it must be acknowledged that it is not wholly without plausibility. Although the object of the 'Tale of a Tub' was undoubtedly to defend the Church of England, and to ridicule its opponents, it would be difficult to find in the whole compass of literature any production more utterly unrestrained by considerations of reverence or decorum. Nothing in Voltaire is more grossly profane than the passages in Swift about the Roman Catholic doctrine concerning the Sacrament, and the Calvinistic doctrine concerning inspiration. And although the 'Tale of a Tub' is an extreme example, the same spirit pervades many of his other performances. His wit was perfectly unbridled. His unrivalled power of ludicrous combination seldom failed to get the better of his prudence; and he found it impossible to resist a jest. It must be added that no writer of the time indulged more habitually in coarse, revolting, and indecent imagery; that he delighted in a strain of ribald abuse peculiarly unbecoming in a clergyman; that he lived in an atmosphere deeply impregnated with scepticism; and that he frequently expressed a strong dislike for his profession. In one of his poems he describes himself as

> A clergyman of special note
> For shunning those of his own coat,
> Which made his brethren of the gown
> Take care betimes to run him down.

In another poem he says:

> A genius in a reverend gown
> Will always keep its owner down;
> 'Tis an unnatural conjunction,
> And spoils the credit of the function.
>
> . . . . .
>
> And as, of old, mathematicians
> Were by the vulgar thought magicians,
> So academic dull ale-drinkers
> Pronounce all men of wit free-thinkers.

At the same time, while it must be admitted that Swift was far from being a model clergyman, it is, I conceive, a complete misapprehension to regard him as an infidel. He was admirably described by St. John as 'a hypocrite reversed.' He disguised as far as possible his religion and his affections, and took a morbid pleasure in parading the harsher features of his nature. If we bear this in mind, the facts of his life seem entirely incompatible with the hypothesis of habitual concealed unbelief. I do not allude merely to the scrupulousness with which he discharged his functions as a clergyman, to his increasing his duties by reading prayers on Wednesdays and Fridays at Laracor, and daily at St. Patrick's, to his administering the Sacrament every week, and paying the most unremitting attention to his choir, and to all other matters connected with his deanery. What I would insist on especially are the many instances of concealed religion that were discovered by his friends. Delany had been weeks in his house before he found out that he had family prayers every morning with his servants. In London he rose early to attend public worship at an hour when he might escape the notice of his friends. Though he was never a rich man he is said to have systematically allotted a third of his income to the poor; and he continued his unostentatious charity when his extreme misanthropy and his extreme avarice must have rendered it peculiarly trying. He was observed in his later years, when it was found necessary to watch him, pursuing his private devotions with the most undeviating regularity; and some of his letters, written under circumstances of agonising sorrow, contain religious expressions of the most touching character.

That he would have been a sceptic if he had not been a clergyman is very probable; but this is no disparagement to his sincerity. It is impossible for any man to throw himself into a profession without the habits, associations, and interests of that profession giving a very real, though perhaps unconscious, bias to his judgment. Few persons can have mixed much with the world without meeting men who are wholly incapable of the hypocrisy of professing what they do not believe, but of whom at the same time it may be safely affirmed that their opinions would have been very different if their judgment had not been in some degree refracted and their natural tendencies checked by professional interests and habits. Swift always flung himself more fully than he allowed into his clerical profession; and, as I have already observed, the advocacy of the High Church theory of government was the constant labour of his life. He employed his own peculiar talent of ridicule continually against the adversaries of his creed, and at least brought the preponderance of wit to the side of orthodoxy; and he never forgot ecclesiastical interests when he was in power. He obtained for the Irish clergy the coveted boon of the remission of the first fruits. The building of fifty new churches in London, under the ministry of Harley, was one of his suggestions. His 'Proposal for the Advancement of Religion,' his admirable letter to a young clergyman on the qualities that are requisite in his profession, the singularly beautiful prayers which he wrote for the use of Stella when she was dying, are all worthy of a high place in religious literature; and although, as he said himself, his sermons were too like pamphlets, they are full of good sense and sound piety admirably and decorously expressed. Of the most political of them—that 'On Doing Good'—Burke has said that it 'contains perhaps the best motives to patriotism that were ever delivered within so small a compass.'

It must be added that the coarseness for which Swift has been so often and so justly censured is not the coarseness of vice. He accumulates images of a kind that most men would have sedulously avoided, but there is nothing sensual in his writings; he never awakens an impure curiosity, or invests guilt with a meretricious charm. His writings in this respect are wholly different from those of Byron, or Sterne, or of French novelists; and it may be safely affirmed that no one has ever been allured to vicious courses by reading them. He is often very repulsive and very indecent, but his faults in this respect are rather those of taste than of morals.

It was not till the year 1713 that Swift's friends succeeded in obtaining for him the Deanery of St. Patrick's. The appointment was regarded both by him and by them as being far below what he might have expected, for its pecuniary value was not great, and it implied separation from all his friends, and residence in a country which was then considered the most unenviable abode for a man of genius. He immediately went over to Ireland, intending to remain there for some time, but was in a few days recalled by his political friends. An open breach had broken out between the Ministers, and the Government seemed on the verge of dissolution. It would be difficult, indeed, to conceive two men less capable of co-operating with cordiality than Harley and St. John, or, to give them the titles they had by this time acquired, than Oxford and Bolingbroke.

Oxford was a man of very moderate abilities and of very unfortunate manners. Frigid, reserved, and formal, he was not popular with any but his most intimate friends, and his fatal habit of procrastination paralysed the energies of the Government. He concealed, however, beneath a cold exterior an affectionate nature; his private life was unusually pure; he showed at different periods of his career an admirable fortitude under adversity and a rare moderation in prosperity; and he was one of the most liberal and enlightened patrons of literature who have ever directed the government of England. Without any of the brilliancy of an orator, or any of the prescience of a great statesman, he maintained, without much difficulty, his position at the head of his party, for he possessed many of the qualities that win confidence in England, and especially among the country gentry and clergy, who constitute the strength of the Tory party. A good private character, moderate views, industry, and business habits weighed more with these classes than the splendid abilities of Bolingbroke, and a certain affectation of mystery, which he often assumed, in some degree enhanced his reputation for wisdom.

His colleague, and at last competitor, was one of the most brilliant and one of the most untrustworthy statesmen who have ever appeared in English public life. The son of a worthless and dissipated character who had fallen in a duel, St. John had been early thrown upon the world, surrounded by all the associations of vice, and endowed by nature with gifts almost as splendid as have ever been united in a single man. With a person of singular beauty, and with a rare charm of manner, he possessed passions so fervid that neither fame nor pleasure could satiate them, and a genius that was equally adapted to sway a senate and to captivate a heart. He plunged with reckless impetuosity into the life of dissipation that opened before him, and, in an age of libertines, was conspicuous as a libertine. Yet even then he found time to amass stores of varied learning and to lay the foundation of those studious habits which were the consolation and the glory of his later years.[1] As an orator he was, by the confession of all his contemporaries, incomparably the foremost of his day, and his writings, though now but little read, are among the

---

[1] In one of the most beautiful of his later Essays he gives us the following sketch of his habits:

'Not only a love of study and a desire of knowledge must have grown up with us, but such an industrious application likewise as requires the whole vigour of the mind to be exerted in the pursuit of truth through long trains of ideas, and all those dark recesses where man, not God, has hid it. This love and this desire I have felt all my life, and I am not quite a stranger to this industry and application. There has been something always ready to whisper in my ear whilst I ran the course of pleasure and of business,—

"Solve senescentem mature sanus equum."

But my genius, unlike the demon of Socrates, whispered so softly, that very often I heard him not in the hurry of those passions by which I was transported. Some calmer moments there were: in them I hearkened to him. Reflection had often its turn, and the love of study and the desire of knowledge never quite abandoned me. I am not, therefore, entirely unprepared for the life I will lead, and it is not without reason that I promise myself more satisfaction in the latter part of it than I ever know in the former.'—*True Use of Retirement and Study.*

most perfect models of English prose. Of the reputation he enjoyed with the best judges of his own generation and of that which immediately followed, it is sufficient to say that the elder Pitt expressed a wish for the recovery of one speech of Bolingbroke rather than any lost work of antiquity; that Chesterfield pronounced his written style equal to that of Cicero, and declared that he would rather his son could attain it than that he should master all the learning of the universities; that Pope made his philosophy the basis of the noblest philosophical poem in the language. Yet notwithstanding his brilliant and varied talents, notwithstanding a great tenacity of purpose, which he displayed chiefly in the latter part of his career, the life of Bolingbroke was in a great degree a failure. In some respects he was singularly unfortunate. The sudden death of the Queen, and, at a later period, the death of the Prince of Wales, baffled his calculations in two of the most critical periods of his life. The indecision and procrastination of Oxford paralysed his energies in one portion of his career, and the bigoted folly of the Pretender consigned him to inactivity in another. But the chief cause of his failure was his own character. It was the restless spirit of intrigue which led him to plot against his colleague, and to enter into relations with the Pretender. It was the notorious dissipation of his private life and the laxity of his opinions, which deprived him of the confidence of his own party and of that of the great majority of the English people.

A rupture between two such statesmen was inevitable. Bolingbroke occupied a position subordinate to Oxford in the Ministry; he had been only created a Viscount when Oxford was created an Earl. His ambition had been perpetually trammelled by Oxford's procrastination, and his consciousness of superior genius irritated by Oxford's haughtiness; and the consequence of all this was, that he conceived a strong dislike to his colleague, which at length deepened into an intense hatred. It is no slight proof of Swift's force of character that he could control two such men, or of the charm of his society that he could retain the affection of both. Personally, he seems to have been especially attached to Oxford; while politically he was inclined to agree with Bolingbroke, that a more energetic line of policy was the only means by which the Tory party could be saved.

In truth, the position of the Government became every week more desperate. The storm of popular indignation, which had been lulled for a time by 'The Conduct of the Allies,' broke out afresh with tenfold vigour on the conclusion of the Peace of Utrecht. The long duration of the war, the numerous Powers engaged in it, and the many complications that had arisen in its progress, rendered the task of the Ministers so peculiarly difficult that it would have been easy to have attacked any peace framed under such circumstances, however consummate the wisdom with which its provisions had been framed. The Peace of Utrecht left England incontestably the first Power of Europe, arrested an expenditure which had been adding rapidly to the national debt, and began one of the most prosperous periods of English history. But, on the other hand, it was undoubtedly negotiated more through party than through national motives; it terminated a long series of splendid victories, and while it saved France from almost complete destruction, it failed to obtain the object for which the war had been begun. The crown of Spain remained upon the head of Philip, and the Catalonians, who had risen to arms relying upon English support, were left without any protection for their local liberties. Any peace which terminated a war of such continual and brilliant success would have been unpopular, and, although the Peace of Utrecht was certainly advantageous to the country, some of the objections to it were real and serious, while its free-trade clauses raised a fierce storm of ignorant or selfish anger among the mercantile classes. Besides this, the Church enthusiasm, which, after the prosecution of Sacheverell, had borne the Tories to power, had begun to subside. The question of dynasty was still uncertain. The leading Tory Ministers were justly suspected of intriguing with the Pretender. They were both, though on different grounds and with different classes, unpopular, and they were profoundly disunited at the very time when their union was most necessary.

Swift, on his arrival from Ireland, induced them to co-operate once more, and he also wrote a defence of the Peace of Utrecht. Having accomplished this, he returned to his deanery, leaving his pamphlet in the hands of the Ministers; but they, being unable to agree about the light in which some transactions connected with the peace were to be represented, withheld the publication, and shortly after quarrelled again. Swift again came to England, but this time his interposition proved unavailing. He then retired from the political scene, and occupied himself in preparing a public Remonstrance addressed to the Ministers, blaming the want of harmony in their councils, and the indecision and procrastination manifest in their actions. Before, however, this Remonstrance was published, the news arrived that Bolingbroke, by the assistance of Lady Masham, had effected the disgrace of Oxford, and had obtained the chief place in the Ministry. Swift received a letter from Lady Masham (who had always been his warm friend), couched in the most affectionate terms, imploring him to continue to uphold the Ministry by his counsel and by his pen, and enclosing an order upon the Treasury for 1000*l.* for the necessary expenses of induction into his deanery, which Oxford had promised, but, with his usual procrastination, had delayed. He received at the same time a letter from Oxford, requesting his presence in the country, where, as the fallen statesman wrote with a touching pathos, he was going 'alone.' Swift did not hesitate for a moment between the claims of friendship and the allurements of ambition; he determined to accompany Oxford.

Events were now succeeding each other with startling rapidity. Bolingbroke had been only four days Prime Minister when the Tory party learned with consternation the death of the Queen, and the consequent downfall of their ascendency. Walpole, who succeeded to the chief power, determined to institute a series of prosecutions for treason against his predecessors. Bolingbroke fled from England, and was condemned while absent. Ormond was impeached. Oxford was thrown into the Tower, where he remained for nearly two years, but was at last tried and acquitted. Swift retired to Ireland. A few vague rumours prevailed of his having been concerned in Jacobite intrigues, but they never took any consistency, or seem to have deserved any attention. 'Dean Swift,' wrote Arbuthnot at this time, 'keeps up his noble spirit, and, though like a man knocked down, you may behold him still with a stern countenance, and aiming a blow at his adversaries.' The misfortunes of his friends, however, and especially the imprisonment of Oxford, profoundly affected him, and he even wrote to the fallen statesman, asking permission to accompany him to prison. He was also at this time, more than once, openly insulted by some Whigs in Dublin, and he had at first serious difficulties with the minor clergy of his deanery.

But a far more serious blow was in store for him—a blow that not only destroyed his peace for a season, but left an indelible stigma on his character. When in London, he had formed a friendship with Miss Vanhomrigh (better known by the name of Vanessa), a young lady of fortune very remarkable for her abilities, though not for her personal beauty. He seems to have been much captivated by her engaging manners and by her brilliant talents; he constantly visited her house, and assisted and directed her in her studies. The possibility of her becoming seriously attached to him appears never for a moment to have flashed through his mind. He had a dangerous fondness for acting the part of monitor or instructor to young ladies of intelligence and grace. He was himself extremely little susceptible to the amatory passion, and, being at this time between forty and fifty, he never seems to have suspected that he could inspire it. He had long been accustomed to a purely intellectual intercourse with Stella, and had probably forgotten how seldom such intercourse retains its first character, and how closely admiration is allied to passion. It was seldom, indeed, that his commanding features—his eye, which Pope described as 'azure as the heavens'—and the charm of his manner and of his wit, failed to exercise a powerful influence on those around him. That spell which had caused Lady Masham to burst into tears when announcing the failure of his ambition; which had controlled Oxford and Bolingbroke in the midst of their dissensions; which had attached to him so many men of genius by a tie that neither his coarseness nor illtemper nor misfortunes could break, acted with a fearful power on his young and enthusiastic pupil. She loved him with all the fervour of an impassioned nature, and an almost adoring reverence blended with and enhanced her affection. The distraction she manifested in her studies betrayed her emotions, and she was compelled to confess her love.

Up to this point the conduct of Swift can hardly be taxed with any graver fault than imprudence, but it now became profoundly culpable. It is evident that he had been much attracted by Vanessa, and the impression she made is curiously shown by the increasing coldness of his Journal to Stella from the early part of 1712, when his acquaintance with her rival began. On the declaration of Vanessa he was filled, as he assures us, with 'shame, disappointment, grief, surprise;' but he shrank with a fatal indecision from the plain and honourable course of decisively severing the connection. He was unwilling to break loose from a companionship he had found so pleasant. He was flattered, as well as surprised, at the passion he had inspired. He miscalculated and misunderstood the force of an affection he had never felt, and having always made a mystery of his connection with Stella, he was probably unwilling to divulge it. A shameful system of temporising was thus begun, which lasted for no less than eleven years. He appears to have attempted, without giving up the connection, to discourage the advances of his pupil, and he probably wrote the poem of 'Cadenus and Vanessa' with that end, though the compliments he paid to her charms must have done much to counteract the effect of his professions of insensibility. When he went to Ireland to his deanery, Vanessa —availing herself of the excuse that she had property in that country—insisted, in spite of Swift's remonstrance, in following him. He cautioned her more than once, and with apparent sincerity, on the imprudence of the step she was taking, but still the friendship was not broken. In the meantime the jealousy of Stella was aroused. It appears to have preyed upon her health, and it inspired her with a beautiful little poem, which is still preserved. Her prior claim was indisputable, and there is very strong evidence that in order to satisfy her a marriage was privately celebrated in 1716. Vanessa continued writing passionate, supplicating letters to Swift, imploring him to marry her. He wrote in reply, sometimes with a coldness of which she bitterly complained. He sometimes assumed an air of repulsion in the interviews he still occasionally had with her. He endeavoured to divert her mind by surrounding her with society, and he openly countenanced a suitor who was seeking her hand; but he never plainly undeceived her, and the strange and somewhat unnatural passion she had conceived for a man of more than fifty continued unwavering and unabated. The death of her sister, leaving her alone in the world, contributed to intensify it. She retired to Celbridge, a secluded country place which she possessed, and there continued to nourish the flame. In letter after letter of feverish impatience she endeavoured to move him, and at length, irritated by his delay, she wrote to Stella. Stella gave the letter to the Dean, who received it with a paroxysm of passion. He rode to Celbridge, entered the room where Vanessa was sitting, and, darting at her a look of concentrated anger, flung down the letter at her feet and departed without uttering a word. She saw at once that her fate was sealed. She languished away, and in a few weeks died. Before her death she revoked the will she had made in favour of Swift, and ordered the publication of 'Cadenus and Vanessa,' the poem in which he had immortalised her love. Swift fled to the country, and remained for two months buried in the most absolute seclusion.

I turn with pleasure from this shameful and melancholy episode to the general tenor of Swift's life in Ireland. The dissensions which had at first existed in his deanery were speedily composed, and he carried on his clerical duties with unremitting energy. He lived in a somewhat parsimonious manner, lodging with a clergyman, but keeping open house twice a week at the deanery. He soon drew around him many acquaintances and a few friends, the principal of whom were Delany, who was one of the fellows of Trinity College, and a schoolmaster named Sheridan, the father of his biographer. Sheridan was in many ways a remarkable character. He was the head of a family which has continued for more than a century to be prolific in genius, having produced a great actor and a great poetess, as well as one of the very greatest of modern orators. He was in many respects a perfect type of the Irish character; recklessly improvident, with boundless good-nature and the most boisterous spirits; full of wit, of fire, and of a certain kind of genius. He ruined his prospects of promotion by preaching from pure forgetfulness from the text 'Sufficient unto the day is the evil thereof' on the anniversary of the accession of the House of Hanover; and all through his life he mismanaged his interests and talents. He carried on a continual warfare with Swift in the shape of puns, charades, satirical poems, and practical jokes; and there is something very winning in the boyish and careless delight with which Swift threw himself into these contests. We owe to them many of his best comic poems, and many of the most amusing anecdotes of his life. It was not to be expected, however, that he could withdraw his attention from political affairs, and he soon entered upon that political career which has given him his place in the history of Ireland.

The position of Ireland was at this time one of the most deplorable that can be conceived. The irreconcileable enmity subsisting between the two sections [1] of the people had issued in the ruin of both parties. The Roman Catholics had been completely prostrated by the battle of the Boyne and by the surrender of Limerick. They had stipulated indeed for religious liberty, but the treaty of Limerick was soon shamelessly violated, and it found no avengers. Sarsfield and his brave companions had abandoned a country where defeat left no opening for their talents, and had joined the Irish Brigade which had been formed in the service of France. They carried with them something of the religious fervour of the old Covenanters, combined with the military enthusiasm so characteristic of Ireland, and they repaid the hospitality of the French by an unflinching and devoted zeal. In the campaign of Savoy, on the walls of Cremona, on the plains of Almanza and of Landen, their courage shone conspicuously. Even at Ramilies and at Blenheim they gained laurels amid the disasters of their friends, while at Fontenoy their charge shattered the victorious column of the English, and is said to have wrung from the English monarch the exclamation, 'Cursed be the laws that deprive me of such subjects!' But while the Irish Roman Catholics abroad found free scope for their ambition in the service of

---

[1] The two religions mark the lines of the antagonism, but do not seem to have been the cause of it. The war was one of races, and not of creeds.

France, those who remained at home had sunk into a condition of utter degradation. All Catholic energy and talent had emigrated to foreign lands, and penal laws of atrocious severity crushed the Catholics who remained. The Protestants were regarded as an English colony; any feeling of independence that appeared among them was sedulously repressed, and their interests were habitually sacrificed to those of England. The Irish Parliament was little more than a court for registering English decrees, for it had no power of passing, or even discussing, any Bill which had not been previously approved and certified under the Great Seal of England. Irishmen were systematically excluded from the most lucrative places. The Viceroys were usually absent for three-fourths of their terms of office. A third of the rents of the country was said to be expended in England, and an abject poverty prevailed. But perhaps the most deplorable characteristic of the time was the complete absence of all public feeling, of all hope, of all healthy interest in political affairs. The Irish nation had as yet known no weapon but the sword. It was broken, and they sank into the apathy of despair.

The commercial and industrial condition of the country was, if possible, more deplorable than its political condition, and was the result of a series of English measures which for deliberate and selfish tyranny could hardly be surpassed. Until the reign of Charles II. the Irish shared the commercial privileges of the English; but as the island had not been really conquered till the reign of Elizabeth, and as its people were till then scarcely removed from barbarism, the progress was necessarily slow. In the early Stuart reigns, however, comparative repose and good government were followed by a sudden rush of prosperity. The land was chiefly pasture, for which it was admirably adapted; the export of live cattle to England was carried on upon a large scale, and it became a chief source of Irish wealth. The English landowners, however, took the alarm. They complained that Irish rivalry in the cattle market was reducing English rents; and accordingly, by an Act which was first passed in 1663, and was made perpetual in 1666, the importation of cattle into England was forbidden.

The effect of a measure of this kind, levelled at the principal article of the commerce of the nation, was necessarily most disastrous. The profound modification which it introduced into the course of Irish industry is sufficiently shown by the estimate of Sir W. Petty, who declares that before this statute three-fourths of the trade of Ireland was with England, but not one-fourth of it since that time. In the very year when this Bill was passed another measure was taken not less fatal to the interests of the country. In the first Navigation Act, Ireland was placed on the same terms as England; but in the Act as amended in 1663 she was omitted, and was thus deprived of the whole colonial trade. With the exception of a very few specified articles, no European merchandise could be imported into the British colonies except directly from England, in ships built in England, and manned chiefly by English sailors. No articles, with a few exceptions, could be brought from the colonies to Europe without being first unladen in England. In 1670 this exclusion of Ireland was confirmed, and in 1696 it was rendered more stringent, for it was enacted that no goods of any sort could be imported directly from the colonies to Ireland. It will be remembered that at this time the chief British colonies were those of America, and that Ireland, by her geographical position, was naturally of all countries most fitted for the American trade.

As far, then, as the colonial trade was concerned, Ireland at this time gained nothing whatever by her connection with England. To other countries, however, her ports were still open, and in time of peace her foreign commerce was unrestricted. When forbidden to export their cattle to England, the Irish turned their land chiefly into sheep-walks, and proceeded energetically to manufacture the wool. Some faint traces of this manufacture may be detected from an early period, and Lord Strafford, when governing Ireland, had mentioned it with a characteristic comment. Speaking of the Irish he says, 'There was little or no manufactures amongst them, but some small beginnings towards a clothing trade, which I had, and so should still discourage all I could, unless otherwise directed by his Majesty and their Lordships. . . . It might be feared they would beat us out of the trade itself by underselling us, which they were well able to do.' With the exception, however, of an abortive effort by this Governor, the Irish wool manufacture was in no degree impeded, and was indeed mentioned with special favour in many Acts of Parliament; and it was in a great degree on the faith of this long-continued legislative sanction that it was so greatly expanded. The poverty of Ireland, the low state of the civilisation of a large proportion of its inhabitants, the effects of the civil wars which had so recently convulsed it, and the exclusion of its products from the English colonies, were doubtless great obstacles to manufacturing enterprise; but, on the other hand, Irish wool was very good, living was cheaper and taxes were lighter than in England, a spirit of real industrial energy began to pervade the country, and a considerable number of

English manufacturers came over to colonise it. There appeared for a time every probability that the Irish would become an industrial nation, and had manufactures arisen, their whole social, political, and economical condition would have been changed. But English jealousy again interposed. By an Act of crushing and unprecedented severity, which was introduced in 1698 and carried in 1699, the export of the Irish woollen manufactures, not only to England, but also to all other countries, was absolutely forbidden.

The effects of this measure were terrible almost beyond conception. The main industry of the country was at a blow completely and irretrievably annihilated. A vast population was thrown into a condition of utter destitution. Several thousands of manufacturers left the country, and carried their skill and enterprise to Germany, France, and Spain. The western and southern districts of Ireland are said to have been nearly depopulated. Emigration to America began on a large scale, and the blow was so severe that long after, a kind of chronic famine prevailed. In 1707 the Irish Government was unable to pay its military establishments, and the national resources were so small that a debt of less than 100,000*l.* caused the gravest anxiety. Fortunately for the country, it was found impossible to guard the ports, and a vast smuggling export of wool to France was carried on, in which all classes participated, and which somewhat alleviated the distress, but contributed powerfully, with other influences, to educate the people in a contempt for law. Industrial enterprise and confidence were utterly destroyed. By a simple act of authority, at a time when the Irish Parliament was not sitting, the English Parliament had suppressed the chief form of Irish commerce, solely and avowedly because it had so succeeded as to appear a formidable competitor; and there was no reason why a similar step should not be taken whenever any other Irish manufacture began to flourish. 'I am sorry to find,' wrote an author in 1729, 'so universal a despondency amongst us in respect to trade. Men of all degrees give up the thought of improving our commerce, and conclude that the restrictions under which we are laid are so insurmountable that any attempt on that head would be vain and fruitless.'[1] Molyneux was impelled, chiefly by these restrictions, to raise the banner of Irish legislative independence. 'Ireland,' wrote Swift, 'is the only kingdom I ever heard or read of, either in ancient or modern story, which was denied the liberty of exporting their native commodities and manufactures wherever they pleased, except to countries at war with their own prince or State. Yet this privilege, by the superiority of mere power, is refused us in the most momentous parts of commerce; besides an Act of Navigation, to which we never assented, pinned down upon us, and rigorously executed.' It may be added that Davenant, who was at this time the chief English writer on economical matters, warmly approved the restriction on Irish wool.

There is one consideration, however, which should not be omitted in estimating the English policy at this period. The intention of Parliament towards Ireland was not purely malevolent, and the address to William in 1698 prayed him to take measures 'for the discouraging the woollen and encouraging the linen manufactures in Ireland,' to which, it was added, 'we shall always be ready to give our utmost assistance.' The reply of the King echoed the address. 'I shall do

[1] An Essay on the Trade of Ireland by the author of 'Seasonable Remarks' (1729).

all,' he said, 'that in me lies to discourage the woollen trade in Ireland, and encourage the linen manufacture and promote the trade of England.' The professed intention of the Legislature was to form a kind of compact, leaving the woollen trade in the possession of England, and the linen trade in that of Ireland.

Upon this compact there are several comments to be made. In the first place it is very obvious to remark that the fact of a nation having created by its industry two forms of manufacture is no possible reason for suppressing one of them; and, as a matter of fact, both had been encouraged by many previous Acts. No one would contend that because the cotton and iron manufactures are both flourishing in England, the destruction of one of them would be other than a fearful calamity. In truth, however, there was no kind of equality between the trade that was permitted and that which was suppressed, and no real reciprocity in the dealings of the two nations. The woollen trade was the chief form of Irish industry. The linen manufacture was as yet so restricted that in 1700 its exports only amounted to a little more than 14,000*l.* The English utterly suppressed the Irish woollen manufacture in order to reserve that manufacture to themselves; but the English and Scotch continued as usual their manufacture of linen. In 1699, when the Irish woollen trade was annihilated, no measure whatever was taken for the benefit of the Irish linen manufacture; and it was not until 1705 that, at the urgent petition of the Irish Parliament, the Irish were allowed to export their white and brown linens, and these only, to the English colonies, but they were not permitted to bring any colonial produce in their return. This concession, which placed one single branch of the linen trade, as far as export to the plantations was concerned, in the position which all Irish goods occupied to the close of the Protectorate, was for many years the sole compensation which England made for the disastrous measure of 1699; and it is a significant fact that it was intended simply for the benefit of the Protestants. The linen trade had been founded, or at least greatly extended, by French Protestant refugees, and had taken root chiefly in the Protestant portion of the island, and the preamble of the Bill for its relief, after reciting the restrictive Act of 1663, proceeds: 'Forasmuch as the Protestant interest of Ireland ought to be supported by giving the utmost encouragement to the linen manufactures of that kingdom, with due regard to her Majesty's good Protestant subjects of her said kingdom, be it enacted,' &c. At a later period, it is true, England was more liberal to this trade. From 1743 bounties were given for its encouragement, which, though never amounting in a single year to much more than 13,000*l*., and usually falling below that amount, were a sign of some solicitude for its interests; but till near the end of the century England reserved for herself a practical monopoly of one branch even of this favoured trade. All dyed or chequered Irish linens were excluded from the colonies till 1777, and were subject to a duty amounting to prohibition if imported to England.[1]

No one, I think, can follow this subject without perceiving how much light it throws upon the later history of Ireland, and upon the character of its people. The successful prosecution of manufacturing industry depends not merely on the accumulation of capital and on natural advantages, but also and quite as much upon the industrial habits of the people, and these are slowly formed by many generations of uninterrupted labour. In England the principal forms of manufacture can be

---

[1] See Hutchinson's 'Commercial Restraints of Ireland.'

traced back in an unbroken history to the time of the Tudors. In Ireland almost every leading industry was checked or annihilated by law, and the linen, which was the only exception, has been successfully developed. The same policy that was pursued with reference to Irish cattle and Irish wool was long afterwards shown in other fields. Thus, to omit many minor and partial restrictions, Ireland was prevented by express enactments or by prohibitory duties from exporting either beer or malt to England, from importing hops from any country but England, from exporting glass (of which she had begun to manufacture the coarser kinds) to any country whatever, from importing it from any country but England.

These last measures, however, belong to a period later than that of Swift. During the time of his Irish career, the management of affairs in Ireland was chiefly in the hands of Archbishop Boulter, who occupied the see of Armagh from 1724 to 1738, and whose correspondence throws much curious and valuable light upon the condition of the country. Boulter was an honest but narrow man, extremely charitable to the poor, and liberal to the extent of warmly advocating the endowment of the Presbyterian clergy; but he was a strenuous supporter of the penal code, and the main object of his policy was to prevent the rise of an Irish party. His letters are chiefly on questions of money and patronage, and it is curious to observe how entirely all religious motives appear to have been absent from his mind in his innumerable recommendations for Church dignities. Personal claims, and above all the fitness of the candidate to carry out the English policy, seem to have been in these cases the only elements considered. His uniform policy was to divide the Irish Catholics and the Irish Protestants, to crush the former by disabling laws, to destroy the independence of the latter by conferring the most lucrative and influential posts upon Englishmen, and thus to make all Irish interests strictly subservient to those of England. The continual burden of his letters is the necessity of sending over Englishmen, to fill all important Irish posts. 'The only way to keep things quiet here,' he writes, 'and make them easy to the Ministry, is by filling the great places with natives of England.' He complains bitterly that only nine of the twenty-two Irish bishops were Englishmen, and urges the Ministers 'gradually to get as many English on the bench here as can decently be sent hither.' On the death of the Chancellor, writing to the Duke of Newcastle, he speaks of 'the uneasiness we are under at the report that a native of this place is like to be made Lord Chancellor.' 'I must request of your Grace,' he adds, 'that you would use your influence to have none but Englishmen put into the great places here for the future.' When a vacancy in the see of Dublin was likely to occur, he writes: 'I am entirely of opinion that the new Archbishop ought to be an Englishman either already on the bench here or in England. As for a native of this country, I can hardly doubt that, whatever his behaviour has been and his promises may be, when he is once in that station he will put himself at the head of the Irish interest in the Church at least, and he will naturally carry with him the college and most of the clergy here.'

It is not surprising that a policy of this kind should have created some opposition among the Irish Protestants, and many traces of dissatisfaction may be found in the letters of Primate Boulter. The Protestants, however, were too few and too dependent upon English support, the Catholics were too prostrate, and public opinion was too feeble and too divided to be very formidable, and measures of the grossest tyranny were carried without resistance, and almost without protest.

There had been, however, one remarkable exception. In 1698, when the measure for destroying the Irish wool trade was under deliberation, Molyneux—one of the members of Trinity College, an eminent man of science, and the 'ingenious friend' mentioned by Locke in his essay—had published his famous 'Case of Ireland,' in which he asserted the full and sole competence of the Irish Parliament to legislate for Ireland. He maintained that the Parliament of Ireland had naturally and anciently all the prerogatives in Ireland which the English Parliament possessed in England, and that the subservience to which it had been reduced was merely due to acts of usurpation. His arguments were chiefly historical, and were those which were afterwards maintained by Flood and Grattan, and which eventually triumphed in 1782. The position and ability of the writer, and the extreme malevolence with which, in commercial matters, English authority was at this time employed, attracted to the work a large measure of attention, and it was written in the most moderate, decorous, and respectful language. The Government, however, took the alarm; by order of the English Parliament, it was burnt by the common hangman, and the spirit it aroused speedily subsided.

Such was the condition of Irish politics and Irish opinion when Swift came over to his deanery. It is not difficult to understand how intolerable it must have been to a man of his character and of his antecedents. Accustomed during several years to exercise a commanding influence upon the policy of the empire, endowed beyond all living men with that kind of literary talent which is most fitted to arouse and direct a great popular movement, and at the same time embittered by disappointment and defeat, it would have been strange if he had remained a passive spectator of the scandalous and yet petty tyranny about him. He had every personal and party motive to stimulate him; he was capable of a very deep and genuine patriotism; and a burning hatred of injustice and oppression was the form which his virtue most naturally assumed.

To this hatred, however, there was one melancholy exception. He was always an ecclesiastic and a High Churchman, imbued with the intolerance of his order. For the Catholics, as such, he did simply nothing. Neither in England when he was guiding the Ministry, nor in Ireland when he was leading the nation, did he make any effort to prevent the infraction of the Treaty of Limerick. He strenuously advocated the Test Act, which excluded the Dissenters from office; and one of his arguments in its favour was, that if it were repealed, even the Catholics, by parity of reasoning, might claim to be enfranchised. The very existence of the Catholic worship in Ireland he hoped would some day be destroyed by law. His language on this subject is explicit and emphatic. 'The Popish priests are all registered, and without permission (which I hope will not be granted) they can have no successors, so that the Protestant clergy will find it perhaps no difficult matter to bring great numbers over to the Church.'

He first turned his attention to the state of Irish manufactures. He published anonymously, in 1720, an admirable pamphlet on the subject, in which he urged the people to meet the restrictions which had been imposed on their trade by abstaining from importation, using exclusively Irish products, and burning everything that came from England—'except the coal.'

He described the recent English policy in an ingenious passage under the guise of the fable of 'Pallas and Arachne.' 'The goddess had heard of one Arachne, a young virgin very famous for spinning and weaving. They both met upon a trial of skill; and Pallas, finding herself almost equalled in her own art, stung with rage and envy, knocked her rival down, turned her into a spider, enjoining her to spin and weave for ever out of her own bowels, and in a very narrow compass.' He concluded with an earnest appeal to the landlords to lighten the rents, which were crushing so many of their tenants. The pamphlet attracted very great attention, but was immediately prosecuted, and Chief Justice Whiteshed displayed the grossest partisanship in endeavouring to intimidate the jury into giving a verdict against it, but the printer ultimately remained unpunished, and a shower of lampoons assailed the judge.

The next productions of Swift were his famous 'Drapier's Letters.' Ireland had been for some time suffering from the want of a sufficiently large copper coinage. Walpole determined to remedy this want, and accordingly gave a person named Wood a patent for coining 108,000*l.* in halfpence. The halfpence were unquestionably wanted, and there is no real ground for believing that they were inferior to the rest of the copper coinage of the country; but there were other reasons why the project was both dangerous and insulting. Though the measure was one profoundly affecting Irish interests, it was taken by the Ministers without consulting the Lord Lieutenant or Irish Privy Council, or the Parliament, or anyone in the country. It was another and a signal proof that Ireland had been reduced to complete subservience to England, and the patent was granted to a private individual by the influence of the Duchess of Kendal, the mistress of the King, and on the stipulation that she should receive a large share of the profits.

It is impossible to justify morally the course which Swift took in this matter, but it may be greatly palliated, especially when we remember that he lived in the age of Bolingbroke and Walpole, when the standard of political morality was far lower than at present. The dignity and independence of the country had been grossly outraged, and an infamous job had been perpetrated, but it would have been hopeless to raise an opposition simply on constitutional grounds. The Catholics were utterly crushed. A large proportion of the Protestants were far too ignorant to care for any mere constitutional question. Public opinion was faint, dispirited, and divided, and the habit of servitude had passed into all classes. The English party, occupying the most important posts, disposing of great emoluments, and controlling the courts of justice, were anxious to suppress every symptom of opposition. The fate of the treatise of Molyneux, and of his own tract on Irish Manufactures, was a sufficient warning, and it was plain that the contemplated measure could only be resisted by a strong national enthusiasm. A report that the coins were below their nominal value had spread through the country, and was adopted by Parliament and embodied in the resolutions of both Houses. Of this report Swift availed himself. Writing in the character of a tradesman, and adopting with consummate skill a style of popular argument consonant to his assumed character, he commenced a series of letters in which he asserted with the utmost assurance that all who took the new coin would lose nearly elevenpence in a shilling, or, as he afterwards maintained with a great parade of accuracy, that thirty-six of them would purchase a quart of twopenny ale. He appealed alternately to every section of the community, pointing out how their special interests would be affected by its introduction, concluding with the beggars, who were assured that the coin selected for adulteration had been halfpence, in order that they too might be ruined. The most terrific panic was soon created. The Ministry endeavoured to allay it by a formal examination of the coin at the Mint, and by a report issued by Sir I. Newton; but the time for such a measure had passed. Swift combated the report in an exceedingly ingenious letter, and the distrust of the people was far too deep to be assuaged.

By this means the needful agitation was produced, and it remained only to turn it into the national channel. This was done by the famous Fourth Letter. Swift began by deploring the general weakness and subserviency of the people. 'Having,' he said, 'already written three letters upon so disagreeable a subject as Mr. Wood and his halfpence, I conceived my task was at an end. But I find that cordials must be frequently applied to weak constitutions, political as well as natural. A people long used to hardships lose by degrees the very notions of liberty; they look upon themselves as creatures of mercy, and that all impositions laid on them by a strong hand are, in the phrase of the report, legal and obligatory.' He defined clearly and boldly the limits of the prerogative of the Crown, maintaining that while the Sovereign had an undoubted right to issue coin he could not compel the people to receive it; and he proceeded to assert the independence of Ireland, and the essential nullity of those measures which had not received the sanction of the Irish Legislature. He avowed his entire adherence to the doctrine of Molyneux; he declared his allegiance to the King, not as King of England, but as King of Ireland; and he asserted that Ireland was rightfully a free nation, which implied that it had the power of self-legislation; for 'government without the consent of the governed is the very definition of slavery.' This letter was sustained by other pamphlets, and by ballads which were sung through the streets, and it brought the agitation to the highest pitch. All parties combined in resistance to the obnoxious patent and in a determination to support the constitutional doctrine. The Chancellor Middleton denounced the coin; the Lords Justices refused to issue an order for its circulation; both Houses of Parliament passed addresses against it; the grand jury of Dublin and the country gentry at most of the quarter sessions condemned it. 'I find,' wrote Primate Boulter, 'by my own and others' enquiry, that the people of every religion, country, and party here are alike set against Wood's halfpence, and that their agreement in this has had a very unhappy influence on the state of this nation, by bringing on intimacies between Papists and Jacobites and the Whigs.' Government was exceedingly alarmed. Walpole had already recalled the Duke of Grafton, whom he described as 'a fair-weather pilot, that did not know how to act when the first storm arose;' but Lord Carteret, who succeeded him as Lord Lieutenant, was equally unable to quell the agitation. A reward of 300*l.* was offered in vain for the discovery of the author of the Fourth Letter. A prosecution was instituted against the printer; but the grand jury refused to find the bill, and persisted in their refusal, notwithstanding the violent and indecorous conduct of Chief Justice Whiteshed. The feeling of the people grew daily stronger, and at last Walpole was compelled to yield and withdraw the patent.

Such were the circumstances of this memorable contest—a contest which has been deservedly placed in the foremost ranks in the annals of Ireland. There is no more momentous epoch in the history of a nation than that in which the voice of the people has first spoken, and spoken with success. It marks the transition from an age of semi-barbarism to an age of civilisation—from the government of force to the government of opinion. Before this time rebellion was the natural issue of every patriotic effort in Ireland. Since then rebellion has been an anachronism and a mistake. The age of Desmond and of O'Neil had passed. The age of Grattan and of O'Connell had begun.

Swift was admirably calculated to be the leader of public opinion in Ireland, from his complete freedom from the characteristic defects of the Irish temperament. His writings exhibit no tendency to exaggeration or bombast; no fallacious images or far-fetched analogies; no tumid phrases in which the expression hangs loosely and inaccurately around the meaning. His style is always clear, keen, nervous, and exact. He delights in the most homely Saxon, in the simplest and most unadorned sentences. His arguments are so plain that the weakest mind can grasp them, yet so logical that it is seldom possible to evade their force. Even his fictions exhibit everywhere his antipathy to vagueness and mystery. As Emerson observes, 'He describes his characters as if for the police-court.' It has been often remarked that his very wit is a species of argument. He starts from one ludicrous conception, such as the existence of minute men, or the suitability of children for food, and he proceeds to examine that conception in every aspect; to follow it out to all its consequences; and to derive from it, systematically and consistently, a train of the most grotesque incidents. He seeks to reduce everything to its most practical form, and to its simplest expression, and sometimes affects not even to understand inflated language. It is curious to observe an Irishman, when addressing the Irish people, laying hold of a careless expression attributed to Walpole—that he would pour the coin down the throats of the nation—and arguing gravely that the difficulties of such a course would be insuperable. This shrewd, practical, unimpassioned tone was especially needed in Ireland. To employ Swift's own image, it was a medicine well suited to correct the weaknesses of the national character.

After the 'Drapier's Letters,' Swift published several minor pieces on Irish affairs, but most of them are very inconsiderable. The principal is his 'Short View of the State of Ireland,' published in 1727, in which he enumerated fourteen causes of a nation's prosperity, and showed in how many of these Ireland was deficient. He also brought forward the condition of the country indirectly, in his amusing proposal for employing children for food—a proposal which a French writer is said to have taken literally, and to have gravely adduced as a proof of the wretched condition of the Irish. His influence with the people, after the 'Drapier's Letters,' was unbounded. Walpole once spoke of having him arrested, and was asked whether he had ten thousand men to spare, for they would be needed for the enterprise. When Serjeant Bettesworth, an eminent lawyer whom Swift had fiercely satirised, threatened him with personal violence, the people voluntarily formed a guard for his protection. When Primate Boulter accused him of exciting the people, he retorted, with scarcely an exaggeration, 'If I were only to lift my finger, you would be torn to pieces.' We have a curious proof of the extent of his reputation in a letter written by Voltaire, then a very young man, requesting him to procure subscriptions in Ireland for the 'Henriade'—a request with which Swift complied, though he had always refused to publish his own works by subscription.

There are few things in the Irish history of the last century more touching than the constancy with which the people clung to their old leader, even at a time when his faculties had wholly decayed; and, notwithstanding his creed, his profession, and his intolerance, the name of Swift was for many generations the most universally popular in Ireland. He first taught the Irish people to rely upon themselves. He led them to victory at a time when long oppression and the expatriation of all the energy of the country had deprived them of every hope. He gave a voice to their mute sufferings, and traced the lines of their future progress. The cause of free trade and the cause of legislative independence never again passed out of the minds of Irishmen, and the non-importation agreement of 1779, and the legislative emancipation of 1782, were the development of his policy. The street ballads which he delighted in writing, the homely, transparent nature of all his pamphlets, and the peculiar vein of rich humour which pervaded them, extended his influence to the very lowest class. It is related of him that he once gave a guinea to a maid-servant to buy a new gown, with the characteristic injunction that it should be of Irish stuff. When he afterwards reproached her with not having complied with his injunction, she brought him his own volumes, which she had purchased, saying they were the best 'Irish stuff' she knew.

But, in spite of all this popularity, Ireland never ceased to be a land of exile to him; and he more than once tried to obtain some English preferment instead of his deanery. With this object, on the death of George I., he made an assiduous court to Mrs. Howard, the mistress of the new Sovereign, but soon found that she possessed no real power. The presence of Pope and Bolingbroke, whom he most truly loved, as well as the wider sphere of ambition it furnished, drew his affections to England, and a number of causes made Ireland peculiarly painful to him. He was engaged towards the close of his life in a multitude of ecclesiastical disputes, into the details of which it is not necessary to enter. He strenuously opposed Bills for commuting the tithes of flax and hemp, for preventing the settlement of landed property on the Church or on public charities, for enlarging the power of the bishops in granting leases, and for relieving pasture land from the payment of tithes; and the first three Bills were ultimately rejected. He was also on very bad terms with the bishops, who were always strong Whigs, and who represented the Church and State policy to which he was most opposed. His judgment of them he expressed with his usual emphasis. 'Excellent and moral men had been selected upon every occasion of vacancy. But it unfortunately has uniformly happened that as these worthy divines crossed Hounslow Heath on their road to Ireland, to take possession of their bishoprics, they have been regularly robbed and murdered by the highwaymen frequenting that common, who seize upon their robes and patents, come over to Ireland, and are consecrated bishops in their stead.'

In 1726 he paid a visit to England, after an absence of twelve years. He was introduced to Walpole, who received him with marked civility, and whom he endeavoured to interest, both directly and through the medium of Peterborough, in Irish affairs. He also revisited his old friends Pope and Bolingbroke, but was soon recalled by the news that Stella was dying. He returned in haste, scarcely expecting to find her alive. 'I have been long weary,' he wrote, 'of the world, and shall, for my small remainder of years, be weary of life, having for ever lost that conversation which could alone make it tolerable.' Stella, however, lingered till 1728. The close of her life was in keeping with the rest, involved in circumstances of mystery and obscurity; and an anecdote is related concerning it which, if it be accepted, would leave a very deep stain on the memory of Swift. The younger Sheridan states, on the authority of his father, that a few days before her death, Stella, in the presence of Sheridan, adjured Swift to acknowledge the marriage that had previously taken place between them, to save her reputation from posthumous slander, and to grant her the consolation of dying his admitted wife. He adds that Swift made no reply, but walked silently out of the room, and never saw her again during the few days that she lived, that she was thrown by his behaviour into unspeakable agonies of disappointment, inveighed bitterly against his cruelty, and then sent for a lawyer and bequeathed her property, in the presence of Sheridan, to charitable purposes. But high as is the authority for this anecdote, there are serious reasons for questioning its accuracy. The book in which it appeared was only published fifty years after the time, and its author was a boy when his father died. It appears from the extant will that it was drawn up, not a 'few days,' but a full month before the death of the testator, and at a time when she was so far from regarding herself as on the point of death that she described herself as in 'tolerable health of body,' left a legacy to one of her servants if he should be alive and in her service at the time of her death, and another to the poor of the parish in which she may happen to die. It is certain that the disposition of her property was no sudden resolution, and it is equally certain that it was not made contrary to the wishes of Swift, for a letter by him exists which was written a year earlier, in which he expresses a strong desire that she could be induced to make her will, and states her intentions about her property in the exact words which she subsequently employed. On money matters, as we have seen, Swift was very disinterested, and it is not surprising that he who had refused to marry Vanessa notwithstanding her large fortune, should have advised Stella to bequeath her property in charity. The terms of agonising sorrow and intense affection in which he at this time wrote about her, and the entire absence of any known reason why he should not have avowed the marriage had she desired it, make the alleged act of harshness very improbable; and it may be added that the will contains a bequest to Swift of a box of papers, and of a bond for thirty pounds. The bulk of her property she beqeathed, as Swift two years before had intimated, to Steevens Hospital, after the death of her mother and sister, to revert to her nearest relative in case of the disestablishment of the Protestant Church in Ireland. It is remarkable that Swift provided for the same contingency in the case of some tithes which he purchased when at Laracor, and left to his descendants. Her body, in accordance with the desire expressed in her will, was buried in St. Patrick's Cathedral.

In addition to the anecdote I have mentioned, there is another related about the last hours of Stella which is not very consistent with the former one. Mrs. Whiteway, the niece of Swift, informed one of his relations that Stella was carried shortly before her death to the deanery, and being very feeble was laid upon a bed, while Swift sat by the side, holding her hand and addressing her in the most affectionate terms. Mrs. Whiteway, out of delicacy, and being unwilling to overhear their conversation, withdrew into another room, but she could not help hearing two broken sentences. Swift said in an audible tone, 'Well, my dear, if you wish it, it shall be owned;' to which Stella answered, with a sigh, 'It is too late;' and it is assumed that these words referred to the marriage. On the whole, there is no decisive evidence that Stella ever complained in her later years of her relations with Swift, or that she suffered from any unhappiness after the death of Miss Vanhomrigh, nor does Swift ever appear during her lifetime to have been accused of harshness to her. The common belief that her death was caused or hastened by unrequited love appears entirely destitute of foundation, and is itself almost absurd. When Stella died she was forty-seven and Swift was sixty-one, and their connection had been unbroken for many years.

It is difficult or impossible to unravel the motives which may have induced Swift to prefer a Platonic marriage to that of ordinary men, but some of them, at least, lie on the surface. He was at first nervously afraid of producing a family upon narrow means; he had in all things a strong bias towards singularity; and he appears to have been absolutely insensible to the passion of love, while he was extremely susceptible to the charms of friendship. These reasons may have at first led to the connection, and the force of habit and the failing health both of himself and of Stella, may have made him unwilling, when he grew richer, to change his habits of life. It is probable, too, as Sir W. Scott has suggested, that some physical cause contributed to his decision. He rarely saw Stella except in presence of a third person, and carefully avoided all occasion of scandal; but she did the honours of his table, though only in the capacity of a guest, on his days of public reception. Her somewhat cold temperament and eminently decorous manners appear to have fallen in well with the arrangement, and there is no evidence of any scandal having been aroused. There is little doubt that she was married to Swift twelve years before her death, but she retained the name of Johnson to the last, and it is still engraven upon her tomb.[1]

But whatever may have been the relation subsisting between Stella and Swift, it is plain that when she died the death-knell of his happiness had struck. 'For my part,' he wrote to one of his friends just before the event took place, 'as I value life very little, so the poor casual remains of it, after such a loss, would be a burden that I most heartily beg God Almighty to enable me to bear; and I think there is not a greater folly than that of entering into too strict and particular a friendship, with the loss of which a man must be absolutely miserable, but especially at an age when it is too late to engage in a new friendship.' That morbid melancholy to which he had ever been subject assumed a darker hue and a more unremitting sway as the shadows began to lengthen upon his path. It had appeared very vividly in 'Gulliver's Travels,' which were published as early as 1726, and which, perhaps, of all his works, exhibits most frequently his idiosyncrasies and his sentiments. We find his old hatred of mathematics displayed in the history of

[1] The Stella mystery has been discussed by all the biographers of Swift, but I must especially acknowledge my obligations to the singularly interesting volume of Dr. Wilde on 'The Last Days of Swift.'

Laputa; his devotion to his disgraced friends, in the attempt to cast ridicule on the evidence on which Atterbury was condemned; his antipathy to Sir Isaac Newton, whose habitual absence of mind is said to have suggested the flappers; as well as allusions to Sir R. Walpole, to the doubtful policy of the Prince of Wales, to the antipathy Queen Anne had conceived against him on account of the indecorous manner in which he had defended the Church, and to a number of other political events of his time. We find, above all, his deep-seated contempt for mankind in his picture of the Yahoos. His view of human nature perhaps differs little from that professed by a large religious school in the present day, but with Swift it was no figure of speech, no mere pulpit dogma, but a deeply realised fact. Living in one of the most hollow, heartless, and sceptical ages that England had ever known, embittered by disappointment and ill-health, and separated by death or by his position from all whom he most deeply loved, he learnt to look with a contempt which is often displayed in 'Gulliver' upon the contests in which so much of his life had been expended, and his naturally stern, gloomy, and foreboding nature darkened into an intense misanthropy. He cast a retrospect over his life, and his deliberate opinion seems to have been that man was hopelessly corrupt, that the evil preponderates over the good, and that life itself is a curse. He appears to have adopted, as far as this world is concerned, the sentiment of his friend Bolingbroke, that there is so much trouble in entering it, and so much in leaving it, that it is scarcely worth while being here at all.

Age had begun to press heavily upon him, and age he had ever regarded as the greatest of human ills. In his picture of the 'Immortals' he had painted its attendant evils as they had never been painted before. He had ridiculed the reverence paid to the old, as resembling that which the vulgar pay to comets, for their beards and their pretensions to foretell the future. He had predicted that, like the blasted tree, he would himself die first at the top. Those whom he had valued the most had almost all preceded him to the tomb. Oxford, Arbuthnot, Peterborough, Gay, Lady Masham, and Rowe, had one by one dropped off. Of all that brilliant company who had surrounded him in the days of his power, Pope and Bolingbroke alone remained; and Pope was sinking under continued illness, and Bolingbroke was drawing his last breath in the more congenial atmosphere of France. Sheridan had gone with broken fortunes to a school at Cavan; Stella had left no successor. His niece, Mrs. Whiteway, watched over him with unwearied kindness, but she could not supply the place of those who had gone.

He looked forward to death without terror and without pain, but his mind quailed at the prospect of the dotage and the decrepitude that precedes it. He had seen the greatest general and the greatest lawyer of his day sink into a second childhood, and he felt that the fate of Marlborough and of Somers would at last be his own. A large mirror once fell to the ground in the room where he was standing. A friend observed how nearly it had killed him. 'Would to God,' he exclaimed, 'that it had!' His mind at length gave way. His flashes of wit became fewer and fewer, and he gradually sank into a condition approaching imbecility, while at the same time his passions became wholly ungovernable. He constantly broke into paroxysms of the wildest fury, into outbursts that were scarcely distinguishable from insanity. Avarice, the common vice of the old, came upon him with a fearful power. He had lost his friends, his talents, and his health, and he clung with desperate tenacity to money, the only thing that remained. He shrank from all hospitality, from all luxuries, from every expense that it was possible to avoid. Yet even at this time he refused a considerable sum which was offered him to renew a lease on terms that would be disadvantageous to his successors.

At length the evil day arrived. A tumour, accompanied by the most excruciating pain, arose over one of his eyes. For a month he never gained a moment of repose. For a week he was with difficulty restrained by force from tearing out his eye. The agony was too great for human endurance. It subsided at last, but his mind had wholly ebbed away. It was not madness; it was absolute idiocy that ensued. He remained passive in the hands of his attendants without speaking, or moving, or betraying the slightest emotion. Once, indeed, when some one spoke of the illuminations by which the people were celebrating the anniversary of his birthday, he muttered, 'It is all folly; they had better leave it alone.' Occasionally he endeavoured to rouse himself from his torpor, but could not find words to form a sentence, and with a deep sigh he relapsed into his former condition. It was not till he had continued in this state for two years that he exchanged the sleep of idiocy for the sleep of death.

He died in 1747, and was buried near the grave of Stella, in his own cathedral, where the following very characteristic epitaph, written by himself, marks his grave:

HIC DEPOSITUM EST CORPUS
JONATHAN SWIFT, S. T. P.
HUJUS ECCLESIÆ CATHEDRALIS
DECANI.
UBI SÆVA INDIGNATIO
COR ULTERIUS LACERARE NEQUIT.
ABI VIATOR,
ET IMITARE SI POTERIS,
STRENUUM PRO VIRILI LIBERTATIS VINDICATOREM.

His property he left to build a madhouse. It would seem as though he were guided in his determination by an anticipation of his own fate. He himself assigned another reason. He says in his poem on his own death :

> He left the little wealth he had
> To build a house for fools and mad,
> To show by one satiric touch
> No nation needed it so much.

The reputation of Swift has suffered from a variety of causes. Politically, he was the founder of an Irish movement which English writers treat, for the most part, with ridicule and contempt, and perhaps the greatest writer of an English party which has steadily been declining. He had also, like so many great men, the misfortune of reckoning among his acquaintances one of those vain and meddling fools who try to win a literary reputation by chronicling the weaknesses of great men. The ' Recollections of Lord Orrery' have furnished materials for much posthumous detraction ; and the extreme coarseness of the writings of Swift, as well as the many repulsive and unamiable features of his character, have given great scope for the censures of the party writer or of the popular moralist.

In truth, the nature of Swift was one of those which neither seek nor obtain the sympathy of ordinary men. Through his whole life his mind was positively diseased, and circumstances singularly galling to a great genius and a sensitive nature combined to aggravate his malady. Educated in poverty and neglect, passing then under the yoke of an uncongenial patron and of an unsuitable profession, condemned during his best years to offices that were little more than menial, consigned after a brief period of triumph to life-long exile in a torpid country, separated from all his friends and baffled in all his projects, he learned to realise the bitterness of great powers with no adequate sphere for their display—of a great genius passed in every walk of worldly ambition by inferior men. His character was softened and improved by prosperity, but it became acrid and virulent in adversity. Hating hypocrisy, he often threw himself into the opposite extreme, and concealed his virtues as other men their vices. Possessing powers of satire perhaps as terrible as have ever been granted to a human being, he employed them sometimes in lashing impostors like Partridge, or arrogant lawyers like Bettesworth, but very often in unworthy personal or political quarrels. He flung himself unreservedly into party warfare, and was often exceedingly unscrupulous about the means he employed; and there is at least one deep stain on his private character; but he was capable of a very genuine patriotism, of an intense hatred of injustice, of splendid acts of generosity, of a most ardent and constant friendship, and it may be truly said that it was those who knew him best who admired him most. He was also absolutely free from those literary jealousies which were so common among his contemporaries, and from the levity and shallowness of thought and character that were so characteristic of his time.

Of the intellectual grandeur of his career it is needless to speak. The chief sustainer of an English Ministry, the most powerful advocate of the Peace of Utrecht, the creator of public opinion in Ireland, he has graven his name indelibly in English history, and his writings, of their own kind, are unique in English literature. It has been the misfortune of Pope to produce a number of imitators, who made his versification so hackneyed that they produced a reaction against his poetry in which it is often most unduly underrated.

Addison, though always read with pleasure, has lost much of his old supremacy. A deeper criticism, a more nervous and stimulating school of political writers have made much that he wrote appear feeble and superficial, and even in his own style it would be possible to produce passages in the writings of Goldsmith and Lamb that might be compared without disadvantage with the best papers of the 'Spectator.' But the position of Swift is unaltered. 'Gulliver' and the 'Tale of a Tub' remain isolated productions, unrivalled, unimitated, and inimitable.

HENRY FLOOD.

The efforts of Swift had created a public opinion in Ireland, but had not provided for its continuance. A splendid example had been given, and the principles of liberty had been triumphantly asserted, but there was no permanent organ to retain and transmit the national sentiment. The Irish Parliament, which seemed specially intended for this purpose, had never been regarded with favour by Swift. He had satirised it bitterly as the Legion Club—

> Not a bowshot from the college,
> Half the world from sense and knowledge;

and its constitution was so defective, and its corruption so great, that satire could scarcely exaggerate its faults. To fire this body with a patriotic enthusiasm, to place it at the head of the national movement, and to make it in a measure the reflex of the national will, was reserved for the subject of the present sketch.

Henry Flood was the son of the Chief Justice of the King's Bench in Ireland. He entered Trinity College as a Fellow-Commoner, but terminated his career, as is still sometimes done, at Oxford. While at the University he applied himself with much energy to the classics, and especially to those studies which are advantageous to an orator in forming a pure and elevated style. For this purpose he learnt considerable portions of Cicero by heart. He wrote out Demosthenes and Æschines on the Crown, two books of the 'Paradise Lost,' a translation of two books of Homer, and the finest passages from every play of Shakespeare. Like most persons who combine great ambition with great powers of expression, he devoted himself much to poetry; his principal production being an 'Ode to Fame,' which was much admired at the time, and is written in the formal, florid style that was then popular. He was also passionately addicted to private theatricals, which were very fashionable, and which contributed not a little to form his style of elocution.

The portraits drawn by his contemporaries are exceedingly attractive. They represent him as genial, frank, and open; endowed with the most brilliant conversational powers, and the happiest manner, 'the most easy and best-tempered man in the world, as well as the most sensible.'[1] His figure was exceedingly graceful, and his countenance, though afterwards soured and distorted by disease, was originally of corresponding beauty. He was of a remarkably social disposition, delighting in witty society and in field-sports, and readily conciliating the affection of all classes. Lord Mountmorres, who knew him chiefly in his later years, and was inclined to judge him with severity, describes him as a preeminently truthful man, and exceedingly averse to flattery. By his marriage he had obtained a large fortune, and was therefore enabled to devote himself exclusively to the service of the country. When we add to this, that he was a man of great eloquence, indomitable courage, and singularly acute judgment, it will be seen that he possessed almost every requisite for a great public leader.

He entered Parliament in 1759 as member for Kilkenny, being then in his 27th year, and took his seat on the benches of the Opposition.

I have said that the Irish Parliament was at this

[1] Grattan.

time subservient and corrupt, and a few facts will show clearly the extent of the evil. The Roman Catholics, who were the vast majority of the population, were excluded from all representation, both direct and indirect. They could not sit in Parliament, and they could not vote for Protestant members. The borough system, which had been chiefly the work of the Stuarts —no less than forty boroughs having been created by James I. alone—had been developed to such an extent that out of the 300 members who composed the Parliament, 216 were returned for boroughs or manors. Of these borough members, 200 were elected by 100 individuals, and nearly 50 by 10. According to a secret report drawn up by the Irish Government for Pitt in 1784, Lord Shannon at that time returned no less than 16 members, the Ponsonby family 14, Lord Hillsborough 9, and the Duke of Leinster 7. An enormous pension list, and the entire patronage of the Government, were systematically and steadily employed in corruption, and this was carried to such an extent that in 1784, besides 44 placemen, the House of Commons contained 86 members who represented constituencies which were let out to the Government in consideration of titles, offices, or pensions. Peerages were the especial reward of borough-owners who returned subservient members, and in this way both Houses were simultaneously corrupted: 53 peers are said to have nominated 123 members of the Lower House.[1] Among the Irish nobility, absenteeism was so common that Swift assures us that in his time the bishops usually constituted nearly half of the working members of the House of Lords;[2] and the ecclesiastical,

[1] See Grattan's Life; Massey's 'History of England;' Lord Cloncurry's 'Recollections.'
[2] Swift's Works (Scott's ed.), vol. viii. p. 365.

like all other appointments, were made chiefly through political motives. At the same time, the House of Commons was almost entirely free from popular control, for, unless dissolved by the will of the Sovereign, it lasted for the whole reign. The Parliament of George II. in this manner continued for no less than thirty-three years.

Any degree of independence that was shown by a body of this kind must have been due chiefly to a conflict between the selfish interests to which it was subject. Collisions between the landlords and the ecclesiastical authorities on the subject of tithes, and between the great Irish nobles and the Government on the subject of patronage, began the independent spirit which the Irish Parliament ultimately showed, and it also, like all legislative bodies, had a natural tendency to extend the sphere of its authority. By a law called Poyning's Law, passed under Henry VII., it had been provided that the Irish Parliament should not be summoned till the Acts it was called upon to pass had been approved under the Great Seal of England, that Parliament could neither originate nor amend any Acts, and that its sole power was that of rejecting the measures thus submitted to it. Gradually, however, these restrictions were relaxed. Parliament regained in a great measure the right of originating Bills, and it claimed, though for a long time unsuccessfully, the right of complete control over the national purse. Its constitutional position before 1782 was a matter of constant dispute between its members and the English authorities, but the prevailing practice is thus described by Lord Mountmorres: Before a Parliament is summoned, he tells us, 'it is necessary that the Lord-Lieutenant and Council should send over an important Bill as the reason for summoning that assembly. This always created violent disputes, and it was constantly rejected, as a money Bill which originated in the Council was contrary to a known maxim, that the Commons hold the purse of the nation. . . . Preparations for laws, or heads of Bills, as they are called, originated indifferently in either House. After two readings and a committal, they were sent by the Council to England, and were submitted usually by the English Privy Council to the Attorney and Solicitor General, and from them they were returned to the Council of Ireland, from which they were sent to the Commons if they originated there (if not, to the Lords), where they went through three stages, and the Lord-Lieutenant gave the royal assent in the same form which is observed in Great Britain. In all these stages in England and Ireland, it is to be remembered that any Bill was liable to be rejected, amended, or altered; but that when they had passed the Great Seal of England, no alteration could be made by the Irish Parliament.'[1] The ultimate form therefore which every Irish measure assumed was determined by the authorities in England, who had the power either of altering or rejecting the Bills of the Irish Parliament, and this latter body, though it might reject the Bill which was returned to it from England in an amended form, had no power to alter it.

The speaking in the Parliament, as might be expected, was in general very bad. Parliamentary eloquence usually implies a certain amount of patriotic enthusiasm, and can scarcely exist when the overwhelming majority are governed by corrupt motives. An eminent lawyer named Malone,[2] who obtained the position of Chancellor of the Exchequer, is said to

[1] Lord Mountmorres's ' History of the Irish Parliament,' vol. i. p. 59.
[2] Father of the well-known editor of Shakespeare.

have been a great master of judicial eloquence; and Grattan, who, in his pamphlet in answer to Lord Clare, has devoted a fine paragraph to him, relates that Lord George Sackville was accustomed to mention him with Chatham and Mansfield as one of the three greatest men he had ever known, but with this exception it appears that before Flood, Ireland had produced no orator of eminence.

Such was the condition of the Parliament when Flood entered upon his career, and made his maiden speech against Primate Stone, who had succeeded to much of the political influence which Boulter had previously possessed, and was the recognised head of the English party. The eloquence and position of the young member soon made him the leader of the party which desired to abridge the corrupt influence of Government, and to establish the independence of Parliament.

His eloquence, as far as we can judge from the description of contemporaries and from the fragments that remain, was not quite equal to that of some later Irish orators. He was too sententious and too laboured. He had, at least in his later years, but little fire and imagination; his taste was by no means pure; and his language, though full of force and meaning, was often tinged with pedantry. He appears, however, to have been one of the very greatest of Parliamentary reasoners. To those who are acquainted with the speeches of Grattan, and know the wonderful force with which that orator condensed an argument into an epigram, and disencumbered it of all superfluous matter, it will be sufficient to say that Flood was invariably considered the more convincing reasoner of the two. He was a great master of grave sarcasm, of invective, of weighty, judicial statement, and of reply;

and he brought to every question a wide range of constitutional knowledge, and a keen and prescient, though somewhat sceptical, judgment. He is also said to have surpassed all his contemporaries in the irritating and embarrassing tactics of an Opposition leader. There was an air of solemn dignity in his manner which added much to the effect of his greater speeches, but did not suit trivial subjects. Grattan said of him, that 'on a small subject he was miserable. Put a distaff into his hand, and, like Hercules, he made sad work of it; but give him a thunderbolt, and he had the arm of a Jove.' The only speaker who was at all able to cope with him in the earlier part of his career was Hely Hutchinson, the Provost of Trinity College,[1] who was superior to him in light sarcasm and raillery, but inferior in all beside.

His indefatigable exertions soon produced their fruit. Public opinion began to show itself outside the walls of Parliament, and a powerful Opposition was organised within. The chief objects he proposed to himself were the shortening of the duration of Parliament, the reduction of the pension list, the creation of a constitutional militia, and the establishment of the principles of Molyneux. In pursuing the first of these objects, he found a powerful auxiliary in Charles Lucas, a very remarkable man who then occupied a prominent position in Irish politics. Lucas had been originally a Dublin apothecary. He was a man of little education and no property, but of a strong, shrewd, coarse intellect, great courage, and indefatigable perseverance. In 1741 he had detected and exposed some encroachments that had been made upon the charters of Irish

---

[1] Author of a most admirable work on the 'Commercial Disabilities of Ireland,' from which I have derived much assistance in that portion of my subject.

corporate towns, and from that time he devoted himself continually to politics. He asserted the independence of Ireland so unequivocally, and he denounced the corruption of Parliament in so pointed and personal a manner, that the grand jury of Dublin at last ordered his addresses to be burnt, and the Parliament, in 1749, proclaimed him an enemy to the country, and issued a warrant for his apprehension. He fled to England, where he became a physician and practised with some success, and he wrote in exile an appeal to the people of both countries, as well as a treatise on Bath waters. A *noli prosequi* at last enabled him to return, and his popularity was so great that he was elected member for Dublin. He had lost the use of his limbs, and his speeches—which were chiefly remarkable for their violent vituperation—were all delivered sitting. He denounced the pensioners and the Government with unsparing bitterness, but there was no one against whom his sarcasm was more envenomed than against his own colleague. That colleague was the Recorder of Dublin, the father of Henry Grattan. Lucas brought forward a Septennial Bill, but it never became law. He assisted Flood in Parliament by his speeches, but exercised a far greater influence outside Parliament by articles in the 'Freeman's Journal,' which he had originated, and which was the foundation of the Irish Liberal press. He died in 1771.[1]

For about ten years the patriotic party in the Irish Parliament carried on a desultory warfare on the questions I have enumerated. Their influence was shown in the creation of a strong and growing public feeling outside Parliament, and of a small but able Opposition within its walls; but though they often embarrassed

---

[1] His pamphlets and addresses have been collected: they form one thick and tedious volume.

a Minister and sometimes carried a division, their measures were always ultimately rejected either by Parliament or the Privy Council. In 1767, however, a great and unforeseen change took place in their prospects, in consequence of the appointment of Lord Townshend as Lord-Lieutenant, and of the new line of policy which he resolved to pursue.

Lord Townshend was brother of the more famous Charles Townshend, whose brilliant but disastrous career closed almost immediately after this appointment. A soldier of some distinction, with considerable talents and popular and convivial manners, he entered upon his administration under very promising circumstances. His first speech favoured the project of making the judges irremovable; and a Bill to that effect was accordingly carried through Parliament, but it was returned from England so altered that it was rejected; and this important reform, which had been obtained in England at the Revolution, was not extended to Ireland till 1782. But the unpopularity which resulted from this failure was more than compensated in the following year by the enthusiasm produced by the concession of one of the strongest wishes of the Irish people. The limitation of the duration of Parliament was justly regarded as the first condition of all constitutional progress, and it was a question upon which a violent agitation had been aroused. The members of Parliament, as was very natural, disliked the change, but they did not venture to resist the popular outcry; they felt secure that if they passed the Bill it would be afterwards rejected in England; and they were not averse to obtaining in this manner some popularity with their constituents. This little comedy was played three times, but in 1768 the English Cabinet resolved to yield. The violent commotion that had arisen in Ireland, the unpopularity produced by the defeat of the Judges Bill, anger at the proceedings of the Irish aristocracy, and perhaps a desire of strengthening the hands of Lord Townshend for the policy he was about to pursue, were their probable motives. The Bill as it passed through Parliament was a septennial one, but was changed in England into an octennial one, and in that form became law. The policy of Flood and Lucas had so far triumphed, and the Parliament became in some real sense an organ of the popular will.

The Lord-Lieutenant, however, who was the object of an enthusiastic ovation in 1768, was destined to become one of the most unpopular who have ever ruled in Ireland, and to give an unprecedented impulse to the national spirit. It had been the custom of his predecessors to reside very little in Ireland, and the management of Parliament was chiefly in the hands of four or five great borough-owners, who undertook to carry on the business of the Government in consideration of obtaining a monopoly of its patronage. This system Lord Townshend resolved to destroy. If his object had been simply to diminish overgrown aristocratic power, to check corruption, or to make Parliament in some degree popular, it would have been laudable, but the real end of his policy appears to have been of a different nature. The great Irish families were grasping, rapacious, and corrupt; but they also constituted in some measure an independent Irish party, and Lord Townshend wished in consequence to break their power, and to make Parliament directly and exclusively subservient to Government influence. With this object, the whole patronage of the Government was employed, and corruption carried to an extent to which even the Irish Parliament was unaccustomed.

The constitutional dependency of the Parliament was strenuously asserted, while the great aristocratic families were thrown into alliance with the party of Flood and of the patriots.

The struggle began upon the question of a money Bill. A large proportion of the Irish members had always, as I have said, aimed at obtaining for their House a complete control of the national purse, and the practice of originating or altering money Bills in England had always been resented. It was contended by some, on very doubtful grounds, that this practice was illegal; by others that, even if strictly legal, it was incompatible with all real national independence, and that Parliament should resist it by the exercise of its undoubted right of rejecting any money Bill which did not originate with itself. A money Bill originated by the Privy Council in 1769 was rejected by the first octennial Parliament on the ground that it did 'not take its rise in that House,' while at the same time the House, to prove its loyalty, voted large supplies to the Crown. The Lord-Lieutenant delivered, in the form of a speech, an angry protest, which he caused to be inserted in the Journals of the House of Lords; and he prorogued the Parliament, though pressing business was on hand. For fourteen months it was not again summoned. In the meantime places were lavishly multiplied. It was afterwards a confession or a boast of Lord Clare that not less than half a million of money was spent in obtaining a majority. With such a constitution as that of the Irish Parliament, such efforts were always in some degree successful. When the House met in 1771, the customary congratulatory addresses to the Lord-Lieutenant were duly carried, though not without great difficulty and after a powerful opposition from Flood in the Commons and from

Charlemont in the Lords; but when another altered money Bill was introduced, it was rejected on the motion of Flood without a division. The Commissioners of Revenue, who were not allowed to sit in the English House of Commons, had seats in that of Ireland, and Lord Townshend, with a view to increasing his Parliamentary influence, resolved to increase their number from seven to twelve. Flood denounced the proposed measure, and on his motion the Parliament passed a resolution asserting the sufficiency of seven. In accordance with another resolution, the opinion of the House was formally laid before the Lord-Lieutenant, who carried out his intention in defiance of Parliament. Every nerve was strained on both sides. A direct vote of censure against those who had advised this increase was then brought forward, and was carried by the casting vote of the Speaker. Lord Townshend succumbed to the storm. He was speedily recalled, but before he left Ireland, he succeeded in obtaining a vote of thanks from Parliament.[1]

During the course of this contest a series of political papers appeared in Dublin, under the title of 'Baratariana,' which produced an extraordinary sensation, and are not even now quite forgotten. They consisted of a history of Barataria, being a sketch of Lord Townshend's administration, with fictitious names; of a series

---

[1] A curious and favourable light is thrown upon the administration of Lord Townshend by some letters of Lord Camden, published in Campbell's 'Lives of the Chancellors,' vol. vi. pp. 386-389. It appears that the Chancellor and the chiefs of the three law courts in Ireland had been always English; that the Irish acquired the King's Bench, and that in the Viceroyalty preceding that of Lord Townshend an Irish Chief Baron was for the first time made. Flood had made a vehement attack upon the plan of sending over judges from England, and Lord Townshend was extremely anxious to make an Irishman Lord Chancellor, but the English Cabinet (guided, as it would appear, chiefly by the advice of Lord Camden and Lord Northington) refused to consent.

of letters modelled after Junius; and of three or four satirical poems. The history and the poems were by Sir Hercules Langrishe, the dedication and the letters signed 'Posthumus' and 'Pertinax' by Grattan, and those signed 'Syndercombe' by Flood. Flood's letters are powerful and well-reasoned, but, like his speeches, too laboured in style, and they certainly give no countenance to the notion started at one time that he was the author of the Letters of Junius.

Flood had now attained to a position that had as yet been unparalleled in Ireland. He had shown that pure patriotism and great abilities could find scope in the Irish Parliament. He had proved himself beyond all comparison the greatest orator that his country had as yet produced, and also a consummate master of Parliamentary tactics. In the midst of a corruption, venality, and subserviency which could scarcely be exaggerated, he had created a party before which Ministers had begun to quail—a party which had wrung from England a concession of inestimable value, which had inoculated the people with the spirit of liberty and of self-reliance, and which promised to expand with the development of public opinion till it had broken every fetter and had recovered every right. No rival had as yet risen to detract from his fame, and no suspicion rested upon his conduct. The tide now began to turn. We have henceforth to describe the rapid decadence of his power. We have to follow him descending from his proud position, eclipsed by a more splendid genius, soured by disappointment, and clouded by suspicion, and sinking, after one brilliant flash of departing glory, into a position of comparative insignificance.

The Administration of Lord Harcourt succeeded that of Lord Townshend. It was conducted on more liberal principles, and Flood at first supported it as an independent member, and at length consented to accept the office of Vice-Treasurer. Of all the steps of his career this has been the most censured, and it is only with great diffidence that I venture to discuss his motives. The materials in print for forming an opinion on this portion of Irish history are so extremely scanty, and they consist in so large a degree of partisan speeches, letters, and biographies, that an historian must always feel painfully conscious that the true springs and motives of the proceedings he describes may lie beyond his knowledge, and that an accurate account of the secret negotiations of the Viceregal Government with the leading statesmen might give a wholly different complexion to his narrative. The reasons, however, which Flood alleged for joining the Government are on record, and, besides contemporary letters and conversations that were preserved, we possess his own very elaborate vindication in a speech which he delivered in 1783 in reply to the invective of Grattan. These reasons seem to me amply sufficient to exculpate him from the charge of corruption. Flood had never been a factious or systematic opponent of Governments, and his persistent hostility to that of Lord Townshend only dated from the prorogation. He desired, it is true, to make the Irish Legislature as independent as that of England, and it was an intelligible policy to stand apart from every Government which refused to make the concession; but such a policy then appeared absolutely suicidal. The constitution of Parliament and the character of its members made it seem utterly impossible that a measure of independence could be carried in the teeth of the Government, and if it were carried there was not the faintest probability of such a movement outside the walls as would compel the English Parliament to yield to it. It was not possible for Flood or for any man to predict the wonderful impulse that was given to the national cause by the American war and by the arms of the volunteers. His success during Lord Townshend's Administration was chiefly due to the accidental alliance of some of the most selfish members of the aristocracy with his party, and even then two votes of thanks to the Lord-Lieutenant were carried in spite of his opposition. When the irritation which Lord Townshend had caused had been allayed by the appointment of a new Viceroy, the party of Flood began at once to dwindle, and it appeared evident that under the existing constitution of Parliament that party could not reasonably hope to do more than modify the course of events. Under these circumstances Flood contended that the true policy of patriots was to act with the Government, and endeavour to make its measures diverge in the direction of public utility. A patriot in office would be obliged to waive the discussion of some measures which he desired, but he could do more for the popular cause than if he were leading a hopeless minority. Flood himself was so indisputably the first man in Parliament that he reasonably held that he could greatly influence the Government, and Lord Harcourt was an honourable and liberal man, and he came to supersede the Viceroy whom Flood had most bitterly opposed. At such a time, and estimating the strength of parties when Ireland was in its normal condition, Flood concluded that the discussion of the independence of Parliament might be advantageously postponed, if its postponement were purchased by some minor concessions on the part of the Government. By becoming Vice-Treasurer he opened to Irishmen an office from which they had been hitherto excluded, he silenced the cry of faction which had been raised against him, and he proved the compatibility of national principles with perfect attachment to the Crown. Ministers had shown themselves willing to make considerable concessions in the direction of economy in order to obtain his support. Some prospect had been held out of a relaxation of the commercial restrictions. They had distinctly authorised him to propose an absentee-tax, to which he, like many Irish Liberals, attached a great importance; and he was not without hopes of being able still further to modify their policy. These reasons, enforced by the persuasive powers of Sir John Blacquiere, determined him, as he said, to accept office, and there appears to me to be no valid reason for questioning his account. It may be added that the faults of his character never were those of corruption. A certain avarice of fame, a nervous solicitude about opinion, made him often jealous of competitors, fretful and uncertain as a colleague, anxious to identify himself with all great measures, and prone to exaggerate his share in their success; but in no other part of his life was he open to a suspicion of being governed by love of money; nor was he in this respect much tempted, for he possessed a large private fortune, and had no children.

Lord Charlemont protested strongly against this resolution of Flood, and there can be no doubt that it formed the fatal turning-point of his life. For nearly seven years he remained in office, and during that period he was obliged to keep resolute silence on those great constitutional questions which in former years he had ceaselessly expounded. His character was no longer above suspicion, and the confidence of the people—the chief element of his power—had passed away. The popular mind always detects readily a change of opinions or of policy, but seldom cares to analyse the motives that may have produced it. The absentee-tax was strongly opposed by the great Whig noblemen in England, and the Government at length abandoned it. The commercial relaxations that he expected were pertinaciously withheld. A two years' embargo was imposed upon Ireland, in consequence of the American war; and in this unpopular measure he was compelled to acquiesce. Like very many politicians of his time, he seems to have regarded the subjugation of America as of vital importance to the empire. 'Destruction,' he once predicted in a characteristic sentence, 'will come upon the British empire like the coldness of death. It will creep upon it from the extreme parts.' Four thousand Irish troops were sent to fight against the Americans. The inducement was, that the pay would be saved to Ireland; the objections were, that it left Ireland without the stipulated number of troops, and in a measure defenceless, and that this extraordinary exertion seemed to imply an extraordinary amount of zeal against a cause which most Liberals regarded as that of justice and of freedom. Flood defended the measure, and designated the troops as 'armed negotiators.' It was to this unfortunate expression that Grattan alluded when he described him, in his famous invective, as standing 'with a metaphor in his mouth and a bribe in his pocket, a champion against the rights of America—the only hope of Ireland, and the only refuge of the liberties of mankind.'

But results such as no one had predicted soon sprang from this measure. The Mayor of Belfast called upon the Government to place a garrison in that town to protect it against the French, and was informed that half a troop of dismounted cavalry and half a troop of invalids were all that could be spared to defend the commercial capital of Ireland.

Then arose one of those movements of enthusiasm

that occur two or three times in the history of a nation. The cry to arms passed through the land, and was speedily responded to by all parties and by all creeds. Beginning among the Protestants of the north, the movement soon spread, though in a less degree, to other parts of the island, and the war of religions and of castes, that had so long divided the people, vanished as a dream. The inertness produced by centuries of oppression was speedily forgotten, and replaced by the consciousness of recovered strength. From Howth to Connemara, from the Giant's Causeway to Cape Clear, the spirit of enthusiasm had passed, and the creation of an army had begun. The military authorities who could not defend the country could not refuse to arm those who had arisen to supply their place. Though the population of Ireland was little more than half of what it is at present, 60,000 men soon assembled, disciplined and appointed as a regular army, fired by the strongest enthusiasm, and moving as a single man. They rose to defend their country alike from the invasion of a foreign army and from the encroachments of an alien Legislature. Faithful to the connection between the two islands, they determined that that connection should rest upon mutual respect and upon essential equality. In the words of one of their own resolutions, 'they knew their duty to their Sovereign, and they were loyal; they knew their duty to themselves, and they were resolved to be free.' They were guided by the chastened wisdom, the unquestioned patriotism, the ready tact of Charlemont. Conspicuous among their colonels was Flood, not uninjured in his reputation by his ministerial career, yet still reverent from the memory of his past achievements and the splendour of his yet unfading intellect; and there, too, was he before whose genius all other Irishmen had begun to pale—the patriot of unsullied purity—the statesman who could fire a nation by his enthusiasm and restrain it by his wisdom—the orator whose burning sentences became the very proverbs of freedom—the gifted, the high-minded Henry Grattan.

It was a moment of supreme danger for the empire. The energies of England were taxed to the utmost by the war, and there could be no reasonable doubt that the Volunteers, supported by the people, could have wrested Ireland from her grasp. A nation unhabituated to freedom, and maddened by centuries of oppression, had suddenly acquired this overwhelming power. Could its leaders restrain it within the limits of moderation? Or, if it was in their power, was it in their will?

The voice of the Volunteers soon spoke, in no equivocal terms, on Irish politics. They resolved that 'Citizens, by learning the use of arms, forfeit none of their civil rights;' and they formed themselves into a regular Convention, with delegates and organisation, for the purpose of discussing the condition of the country. Their denunciations of the commercial and legislative restrictions grew louder and louder; and two cannons were shown labelled with the inscription 'Free Trade or this!'

In Parliament Grattan and Hussey Burgh made themselves the interpreters of the prevailing feeling. The latter, in a speech which was long remembered as a masterpiece of eloquence, described the condition of the country, and called upon the Ministers to avert war by timely and ample concessions. 'Talk not to me,' he exclaimed, 'of peace; it is not peace, but smothered war. England has sown her laws in dragons' teeth, and they have sprung up in armed men.' The restrictions on trade were made the special objects of attack. I have described in the last chapter the manner in which—with the exception of the linen trade—almost every branch of Irish commerce and manufacture was crippled or ruined by law, and very few measures of relief had been carried during the first three quarters of the eighteenth century. Some additional encouragement had indeed been given to Irish linen. Several temporary Acts were passed permitting Irish cattle, salted provisions, and tallow to enter England, and in 1765 Ireland was allowed to receive iron and timber direct from the colonies, but the more important disabilities remained unchanged. In 1775, however, a strong movement for free trade arose in Ireland, which fully triumphed under the influence of the Volunteers in 1779. In the first of these years Irish vessels were admitted to the fisheries of Newfoundland and Greenland. In 1778 several small relaxations were made in the prohibitory laws which excluded Ireland from the colonial trade. In the beginning of 1779 an attempt was made to allay the Irish cry for the repeal of all commercial disabilities by granting new bounties to linen and to hemp, and by permitting the cultivation of tobacco in Ireland. The time, however, for such compromise had passed, and on both sides of the Channel public feeling ran dangerously high. The English manufacturers, and especially the towns of Manchester and Glasgow, were bitterly opposed to any measure of free trade, and their opposition hampered the very liberal tendencies of Lord North. The Irish were in arms, and they demanded nothing less than to be placed on the same footing with the English. Numerous meetings were held, and resolutions adopted, pledging the people neither to import or consume any articles of English manufacture till the commercial restrictions were removed; and when Parliament met in October 1779,

Burgh moved, as an amendment to the address from the throne, a petition for 'an extension of trade.' Flood, who was still a Minister, rose and suggested that the expression 'free trade' should be employed, and spoke in favour of the amendment, which was carried. The House went in a body to present their petition to the Lord-Lieutenant, and the Volunteers lined the road and presented arms to them as they passed. The due emphasis was thus supplied to their request, and Lord North soon after brought forward in England a series of measures which removed the chief grievances that were complained of. The Acts prohibiting the Irish from exporting their woollen and glass manufactures were repealed, and the colonial trade was thrown open to Ireland.

The events that have been described rendered the position of Flood as Minister still more irksome than it had been, and at last he took the step which it was plainly his duty to have taken before—threw up his office and rejoined his old friends. The Ministers marked their displeasure at his conduct by dismissing him from the Council; and he never regained his former position among the Liberals in Parliament. He found that his long services had been forgotten during his long silence, that the genius of Grattan had obtained a complete ascendency in Parliament, and that the questions he had for so many years discussed were taken out of his hands. He felt the change very acutely, and it exercised a perceptible influence upon his temper. In 1779 Yelverton brought forward a Bill for the repeal of Poyning's Law; and Flood, while supporting the measure, complained bitterly that 'after a service of twenty years in the study of this particular question' he had been superseded. He added: 'The honourable gentleman is erecting a temple of liberty. I hope that at least I shall be allowed a niche in the fane.' Yelverton retorted by reminding them that by the civil law ' if a man should separate from his wife, desert, and abandon her for seven years, another might then take her and give her his protection.'

I pass over the events that immediately followed the discussions of the Volunteers, and the ultimate triumph of Irish independence, as belonging more especially to the life of Grattan. The next prominent transaction in which Flood appears was the fatal controversy on the subject of Simple Repeal. How far in this matter he was actuated by personal motives, and how far by pure patriotism, it is impossible to determine. This much may be said in his favour—that he supported every step of his policy by specious if not by conclusive arguments, and that he carried with him a very large section of the intellect of the country. The broad question on which he differed from Grattan was the advisability of continuing the Volunteer Convention. Grattan wished Ireland to subside into its normal condition as soon as the independence of the Parliament had been declared; he felt the danger and the irregularity of having the representatives of an armed force organised like an independent Parliament, and overawing all other authority in the land. He considered that Parliamentary reforms should emanate from Parliament alone, and should be the result of no coercion, except that of public opinion. Flood, on the other hand, perceived that Ireland was in a position, with reference to England, such as she might never occupy again; he believed that by continuing the Convention a little longer, guarantees of Irish independence might be obtained which it would be impossible afterwards to overthrow; and that Parliament might be so reformed as to be made completely subject to public opinion, and therefore completely above the danger of ministerial intrigue. He foresaw what Grattan at that time does not appear to have foreseen, that the English Ministers would never cordially accept the new position of Ireland; that they would avail themselves of every extraordinary circumstance, of every means of corruption in their power, to strangle the independence of Parliament; and that the borough system gave them a fatal facility for the accomplishment of their purpose.

The Simple Repeal controversy may be thus shortly stated: English statesmen maintained, and Irish Liberals, from Molyneux to Grattan, denied, that the effect of Poyning's Law was to make the Irish Parliament entirely subservient to English control.[1] The Parliament of England fixed the sense by a declaratory Act, asserting the dependence of that of Ireland, and it was on these two enactments that its authority in Ireland rested. In 1782 the Irish Parliament asserted its own independence, and the English Parliament repealed its declaratory Act. The question at issue was whether this was sufficient, or whether an express renunciation should be exacted from England.

Grattan argued that the principle of dependence was embodied in the declaratory Act, and therefore that its repeal was a resignation of the pretended right; that when a man of honour affirms that he possesses a

---

[1] The following is Bacon's account of its origin and nature: 'Poyning, the better to make compensation of the meagreness of his services in the wars by acts of peace, called a Parliament, when was made that memorable Act which at this day is called Poyning's Law, whereby all the statutes of England were made to be of force in Ireland, for before they were not; neither are any now in force in Ireland which were made since that time, which was the eighteenth year of the King.' —*History of Henry VII.*

certain power, and afterwards solemnly retracts his declaration, it is equivalent to a distinct disavowal, and that the same laws of honour apply to nations and to individuals; that to require an express renunciation from England would be to exhibit a distrustful and an overbearing spirit, and would keep alive the ill-feeling between the two countries which it was most important to allay; that it would also stultify the Irish Liberals, for it would imply that England actually possessed the right she was called upon to renounce.

To these reasonings it was replied that the declaratory law did not make a right, and that therefore its repeal could not unmake it; that though Irish Liberals maintained that England had never possessed the right in question, the English Parliament had asserted its authority, and that the repeal of the declaratory Act was not necessarily anything more than the withdrawal of that assertion as a matter of expediency for the present; that an express renunciation would be a charter of Irish liberties such as no legal quibble could evade; and that if England had no desire to re-assert her claim, she could have no objection to make it. It was added that the history of English dealings with Ireland showed plainly how necessary it was to leave no loophole or possibility of encroachment. It was a peculiarity of the Irish question that the independence of the Irish Parliament was bitterly opposed in England, on different grounds, by the most opposite parties. The high prerogative party objected to it as a measure of political emancipation. The trading classes, who constituted the chief strength of the Whig party, were equally opposed to it through their jealousy of Irish trade.

In addition to these general considerations, several

circumstances had occurred in England which greatly disturbed the public mind. Lord Abingdon, in the English House of Lords, had drawn a distinction between a right to internal and a right to external legislation, and had argued that, while England had relinquished the former, she had retained the latter. An English law with reference to the importation of sugar from St. Domingo had been drawn up in terms that seemed applicable to Ireland, and Lord Mansfield had decided an old Irish law case.

The Simple Repeal question was not started by Flood, but it gained its importance chiefly from his adhesion to the party who were yet unsatisfied. He brought forward their arguments with his usual force, and concluded his speech with an appeal of great solemnity, which bears every mark of earnest feeling. 'Were the voice,' he said, 'with which I now utter this, the last effort of expiring nature; were the accent which conveys it to you the breath that was to waft me to that grave to which we all tend, and to which my footsteps rapidly accelerate, I would go on, I would make my exit by a loud demand for your rights: and I call upon the God of truth and liberty, who has so often favoured you, and who has of late looked down upon you with such a peculiar grace and glory of protection, to continue to you His inspirings, to crown you with the spirit of His completion, and to assist you against the errors of those that are honest, as well as against the machinations of those that are not.' Most of the Volunteers, headed by the lawyer corps, whose opinion on such a question naturally carried great weight, supported Flood, and the popularity of Grattan in the country waned as rapidly as it had risen. It became customary to say that nothing had really been gained until the formal renunciation had been made; and at last Fox brought forward in England the required renunciatory Act.

It was in the course of this controversy that the famous collision between Flood and Grattan took place. It had been for some time evident to close observers that it must come sooner or later. For several years the friendship between these two great men had been growing colder and colder, and giving way to feelings of hostility. Flood felt keenly the manner in which he had been superseded as leader of the Liberals. He could not reconcile himself to occupying a second place to a man so much younger than himself, after having been for so long a period the most conspicuous character in the country. The particular subject of the independence of Parliament he had brought forward again and again when Grattan was a mere boy, and it seemed hard that another should reap the glory of his long and thankless labour. He had sat in Parliament for sixteen years before Grattan had entered it. He had borne the brunt of the battle at a time when the prospects of the cause seemed hopeless; and if less brilliant than his rival he was deemed by most men fully his equal in solid capacity, and greatly his superior in experience. Grattan, on the other hand, regarded Flood's adhesion to the Harcourt Administration as an act of apostacy, and his agitation of Simple Repeal as a struggle for a personal triumph at the expense of the interests of the country. He dreaded the permanence of the Volunteer Convention, the increase of the ill-feeling existing between the two countries, and a needless and dangerous agitation of the public mind. Ill health and the position he had so long held had given Flood a somewhat authoritative and petulant tone, which contrasted remarkably with his urbanity in private life; and Grattan, on his side, was embittered by the sudden decay of his popularity, and by several slight and not very successful conflicts with his rival.

Under these circumstances it needed but little to produce an explosion, and that little was supplied by a singularly discourteous and unfair allusion to Flood's illness which escaped from Grattan in the heat of the debate. Flood rose indignantly, and, after a few words of preface, launched into a fierce diatribe against his opponent. His task was a difficult one, for few men presented a more unassailable character. Invective, however, of the most outrageous description, was the custom of the time, and invective between good and great men is necessarily unjust. He dwelt with bitter emphasis on the grant the Parliament had made to Grattan. He described him as 'that mendicant patriot who was bought by his country, and sold that country for prompt payment;' and he dilated with the keenest sarcasm upon the decline of his popularity. He concluded, in a somewhat exultant tone: 'Permit me to say that if the honourable gentleman often provokes such contests as this, he will have but little to boast of at the end of the session.' Grattan, however, was not unprepared. He had long foreseen the collision, and had embodied all his angry feelings in one elaborate speech. Employing the common artifice of an imaginary character, he painted the whole career of his opponent in the blackest colours, condensed in a few masterly sentences all the charges that had ever been brought against him, and sat down, having delivered an invective which, for concentrated and crushing power, is almost or altogether unrivalled in modern oratory.

Thus terminated the friendship between two men

who had done more than any who were then living for their country, who had known each other for twenty years, and whose lives are imperishably associated in history. Flood afterwards presided at a meeting of the Volunteers, where a resolution complimentary to Grattan was passed; Grattan, in his pamphlet on the Union, and more than once in private conversation, gave noble testimony to the greatness of Flood; but they were never reconciled again, and their cordial co-operation, which was of such inestimable importance to the country, was henceforth almost an impossibility.

The dissension between the Parliament and the Volunteers had now become very marked, and it was evident that there existed among the latter a party who desired open war with England. It is curious that their leader should have been by birth an Englishman, and by position a bishop. The Earl of Bristol and Bishop of Derry was son of that Lord Hervey who was long remembered only as the object of the fiercest of all the satires of Pope, but who within the last few years has been revealed in altogether a new light, by the publication of those masterly memoirs in which he had described the court and much of the State policy of George II. The character of the Bishop has been very differently painted, but its chief ingredients are sufficiently evident, whatever controversy there may be about the proportions in which they were mixed. He appears to have been a man of respectable learning and of real talent, sincerely attached to his adopted country, and on questions of religious disqualification greatly in advance of most of his contemporaries; but he was at the same time utterly destitute of the distinctive virtues of a clergyman, and he was one of the most dangerous politicians of his time.

Vain, impetuous, and delighting in display, with an insatiable appetite for popularity, and utterly reckless about the consequences of his acts, he exhibited, though an English peer and an Irish bishop, all the characteristics of the most irresponsible adventurer. Under other circumstances he might have been capable of the policy of an Alberoni. In Ireland, for a short time, he rode upon the crest of the wave; and if he had obtained the control he aspired to over the Volunteer movement, he would probably have headed a civil war. But though a man of clear, prompt judgment, of indisputable courage, and of considerable popular talents, he had neither the caution of a great rebel nor the settled principles of a great statesman. His habits were extremely convivial; he talked with reckless folly to his friends, and even to British officers, of the appeal to arms which he meditated; and he exhibited a passion for ostentation which led men seriously to question his sanity. 'He appeared always,' says Barrington, 'dressed with peculiar care and neatness, generally entirely in purple, and he wore diamond knee and shoe buckles; but what I most observed was, that he wore white gloves with gold fringe round the wrists, and large gold tassels hanging from them.' The ostentation he manifested in his dress he displayed in every part of his public life. A troop of horse, commanded by his nephew, used to accompany him when he went out, and to mount guard at his door. On one occasion he drove in royal state to a great meeting which was held at the Rotundo, escorted by a body of the Volunteers, who sounded their trumpets as they passed the Parliament-house, much to the astonishment of the assembled members.

Fortunately, however, the influence of the Bishop with the Volunteers, though very great, was not absolute. He desired to become their president, but, though he had many partisans, Lord Charlemont was elected to the place; and in the Convention itself the practised oratory of Flood gave him a complete ascendency. At the same time, it is not surprising that the proceedings of the Volunteers should have created much alarm in many minds, and that strong wishes should be felt for the dissolution of the Convention. But for this measure Flood was not prepared. He maintained that two great dangers had menaced the independence of Parliament, that it might be evaded by a legal quibble, and that it might be betrayed by the corruption of its members. By obtaining from England a distinct renunciation of all supremacy, he had provided effectually against the first of these dangers. By reforming the Parliament, he sought to guard against the latter. But, in order that a Reform Bill should be brought forward with any chance of success, he believed it to be essential that it should be supported by all the threatening weight of the Volunteer Convention. Had he succeeded in carrying the reform he meditated, he would have placed the independence of Ireland on the broad basis of the people's will, he would have fortified and completed the glorious work that he had himself begun, and he would have averted a series of calamities which have not even yet spent their force. We should then never have known the long night of corruption that overcast the splendour of Irish liberty. The blood of 1798 might never have flowed. The Legislative Union would never have been consummated, or, if there had been a Union, it would have been effected by the will of the people, and not by the treachery of their representatives, and it would have been remembered only with gratitude or with content.

The Reform Bill was drawn up by Flood, and was first submitted to the Volunteer Convention for their sanction. In one respect it was glaringly defective. It proposed to extend the franchise largely, but it gave no political power to the Catholics. On this point both Flood and Charlemont were strenuously opposed to Grattan; and when, in 1782, a measure had been brought forward to enable the Catholics to purchase estates, Flood strongly supported an amendment excepting all borough rights by which members might be returned to Parliament. With this grave exception, the measure was a comprehensive one, and would have effectually cured the great evils of the Legislature. It proposed to open the close boroughs by giving votes to all Protestant forty-shilling freeholders, and to leaseholders of thirty-one years, of which fifteen were unexpired. It provided that in the case of decayed boroughs the franchise should be extended to the adjoining parishes; that pensioners who held their pensions during pleasure should be excluded from Parliament; that those who accepted a pension for life or a Government place should vacate their seats; that each member should take an oath that he had not been guilty of bribery at his election; and that the duration of Parliament should be limited to three years.

It was in truth a night of momentous importance to the country when Flood brought forward in Parliament the Volunteer Reform Bill, and the crowded benches and the anxious faces that surrounded him showed how fully the magnitude of the struggle was appreciated. The elation of recovered popularity and the proud consciousness of the grandeur of his position, dispelled the clouds that had so long hung over his mind, and imparted a glow to his eloquence worthy of his brightest days. He had too much tact even to mention the Volunteers in his opening speech; but the uniform he wore, the fire of his eye, and the almost regal majesty of his tone and of his gesture reminded all who heard him of the source of his inspiration. He was opposed by Yelverton, the Attorney-General. Yelverton was at all times a powerful speaker, but on this night he seems to have made his greatest effort. He called upon the House to reject the Bill without even examining its intrinsic merits, as coming from the emissaries of an armed body; he denounced it as an insult and a menace, as a manifest infringement of the privileges of Parliament; and he appealed to all parties to rally round the liberties of their country, so lately rescued from English domination, and now threatened by a military council. Flood, in his reply, rested—perhaps rather disingenuously—on his not having spoken of the Volunteers. He had not mentioned them, but if they were attacked he was prepared to support them; and then he digressed, with the adroitness of a practised debater, into their defence. He reminded his hearers how much they owed to that body; how the Volunteers had emancipated their trade and struck off their chains; how absurd, how ungrateful it would be to assail their deliverers as enemies, and to brand them as hostile to liberty. Yet it was not for the Volunteers that he asked reform; he would rather place the question on its own merits. 'We come to you,' he said, 'as members of this House; in that capacity we present you with a Reform Bill. Will you receive it from us?'

He was, however, but feebly supported and strongly opposed. Many members dreaded reform on personal grounds, and were doubtless glad of a plausible pretext for opposing it; others believed that the Convention was the most pressing danger. Lord Charlemont, the leader of the Volunteers, who, though not a member, had a great influence in the Lower House, was timid, vacillating, and perplexed. The Government exerted all its influence against Flood, and a majority, actuated by various motives, rejected the Bill. The numbers were 158 to 49, and it is said that more than half the majority were placemen. A resolution to the effect that the dignity of the House required asserting, which was tantamount to a censure of the Volunteers, was then moved and carried. Grattan voted with Flood on the reform question, and against him on the subsequent resolution. Lord Charlemont adjourned the Convention *sine die*, and its members separated with an alacrity and a submission that furnished the most eloquent refutation of the charges of their opponents.

The conduct of Flood in this transaction has given rise to much controversy, and it is difficult to pronounce very decidedly upon it. There can be no question that the existence of an assembly consisting of the representatives of a powerful military force, convened for the purpose of discussing political questions, was extremely menacing, both to the Parliament and the connection. If the Bishop of Derry had obtained the presidency, matters would probably have been pushed to a rebellion. This period was perhaps the only one in Irish history when the connection between the two countries might have been easily dissolved, and when the dissolution would not have involved Ireland in anarchy or civil war. In the prostrate condition to which England had been reduced, she could scarcely have resisted an organised army, which rose at last to more than 100,000 soldiers, which was commanded by the men of most property and influence in the country, and was supported by the enthusiasm of the nation. Such an organisation was far more powerful than that which had just wrested the colonies from her grasp. Had the severance been effected, Ireland possessed a greater amount of legislative talent than at any former period, and her newly emancipated Parliament only needed a reform to become a most efficient organ of national representation. There was then no serious conflict of classes, and the Catholic question, though it caused division among politicians, was at this time no source of danger to the country. The Catholics had neither education, leaders, nor ambition. They were perfectly peaceful, and indeed quiescent, and the process of emancipation would probably have been carried out silently and tranquilly. The most obnoxious of the penal laws had already been repealed. The Volunteers had passed a resolution approving of that repeal. The rising school of politicians were in favour of granting political power to the Catholics, and the cause had no more unhesitating supporter than the Bishop of Derry.

This was the course which the Volunteer movement would probably have taken if the influence of the Bishop had prevailed. Flood, however, does not appear to have had any desire to produce rebellion,[1] and he was no friend of Catholic emancipation. His object was to overawe the Parliament by the menace of military force, in order to induce it to reform itself. It is sufficiently manifest that such an attempt was extremely dangerous and unconstitutional, but it was a desperate remedy applied to a desperate disease. It

---

[1] See, however, on the other side, a curious traditionary anecdote related by O'Connell, on the authority of Bartholomew Hoare, a friend of Flood, and preserved in O'Neil Daunt's 'Ireland and her Agitators,' pp. 4, 5.

was a matter of life or death to the Irish Constitution that the system of corruption and rotten boroughs which gave the Castle a sure and overwhelming majority should be ended, and, as a great majority of the members had a personal interest in its permanence, some degree of intimidation was absolutely necessary. Even the Reform Bill of 1832 would never have been passed if the country had been tranquil. There was, no doubt, a considerable difference between the display of force to carry free trade and legislative independence in accordance with the wishes of Parliament, and the display of a similar force to overawe the Parliament; but if the liberties of Ireland were to be permanent, the reform was absolutely necessary, and at this time it could in no other way have been effected. Had Charlemont, Grattan, and Flood been cordially united, it would probably have been forced through Parliament, and the Constitution of 1782 would have been established. Whether, however, the Volunteers, flushed with a new conquest, would have consented to disband, may reasonably be doubted. Fox, in a very earnest letter, urging the Irish Government to resist the Volunteer demand to the uttermost, said: 'The question is not whether this or that measure shall take place, but whether the Constitution of Ireland, which Irish patriots are so proud of having established, shall exist, or whether the Government shall be as purely military as ever it was under the Prætorian bands.' The defensive utility of the Volunteers had terminated with the peace; and their desire of encroaching on the political sphere had grown. I venture, however, to think that the probabilities were, on the whole, in favour of the peaceful dispersion of the force when its work was accomplished. The French Revolution, which has given so violent and democratic a tendency to most popular movements, had not yet taken place. The Volunteers, as I have said, were guided by the rank and property of the country, and these were amply represented in the Convention. Above all, the moderation of the assembly in selecting Charlemont for its head, and in dispersing peacefully after its defeat, may be taken as a sufficient evidence of the patriotism of its members.

All the leading men, however, were somewhat below the occasion. Grattan was not a member of the Convention. He would not co-operate with Flood, and he utterly disapproved of the continuance of the Convention, and of all attempts to overawe the Legislature. Charlemont remained at the head of the Volunteers chiefly in order to moderate them, and his opinion on the question at issue was, in reality, little different from that of Grattan. The Bishop of Derry was violent, vain, and foolish. Flood was but too open to the imputation of having stirred up the question of simple repeal through envy at the triumph of Grattan, and of aggrandising the power of the Convention, in which he was almost supreme, through jealousy of Parliament, in which his influence had diminished. In undertaking an enterprise of so perilous and unconstitutional a character, it ought at least to have been made certain that the voice of the people was with the Volunteers; but no step whatever appears to have been taken to obtain petitions or demonstrations, and at the very time when Flood was pushing the country to the verge of a civil war, he was damping the enthusiasm of the Catholics by carefully excluding them from his scheme of reform.

The effects of this episode upon the country were very injurious. Violent riots broke out in Dublin, and the mob forced its way into the Parliament House.

The Parliament had shown some spirit in refusing even to entertain a Bill emanating from a military force, but, as it refused with equal pertinacity to yield to subsequent Reform Bills which were brought forward without military assistance, and with the support of petitions from twenty-six counties, it neither received nor deserved credit. The Volunteer Convention dissolved itself; but the Volunteers themselves, with diminished importance, and under the guidance of inferior men, continued for many years in a divided and broken state, and the United Irishmen rose out of their embers.

The bishop who had occupied so prominent a place in the movement afterwards retired, on the plea of ill-health, to Italy, where he lived for many years a wild and scandalous life, retaining the emoluments but utterly neglecting the duties of his bishopric, scoffing openly at religion, and adopting without disguise the lax moral habits of Neapolitan society. His wealth, his good-nature, his munificent patronage of art,[1] and his brilliant social qualities made him very popular, and in his old age he was a lover of Lady Hamilton, to whom he was accustomed to write in a strain of most unepiscopal fervour. He fell into the hands of the French in 1799, and was imprisoned at Milan for eighteen months. He died near Rome in 1803.[2]

The career of Flood in the Irish Parliament was

---

[1] We have an amusing illustration of his art taste in an engraving of one of the most indecent of the pictures of Albano. 'Actæon Discovering Diana and her Nymphs just Emerging from the Bath,' which is dedicated to 'the Earl of Bristol and Lord Bishop of Derry;' underneath are the bishop's arms surmounted by the mitre, and a little below the mitre is the bishop's motto—'Je ne l'oublierai jamais.'

[2] There is much curious information about the latter years of this eccentric bishop in the 'Memoirs of Lady Hamilton,' and in those of the Comtesse de Lichtenau.

now rapidly drawing to a close. In the following year he made another effort to induce the Parliament to reform its constitution; but, as he was doubtless well aware, such an attempt, when opposed by the Government and unsupported by the Volunteers, was at that time almost hopeless. The Reform Bill, notwithstanding the petitions in its favour, was rejected, and Flood shortly after put into execution a design that he had conceived many years before, of entering the Parliament of England. His failure there is well known. His habits had been already formed for an Irish audience, and, as Grattan said of him, 'he was an oak of the forest too great and too old to be transplanted at fifty.' He was also guilty of much imprudence. Desiring to act in the most independent manner, he proclaimed openly that he would not identify himself with either of the great parties in Parliament. He thus prejudiced both sides of the House against him, and deprived himself of that support which is of such great consequence to a debater. He spoke first on the India Bill, which ultimately led to the downfall of the Coalition Ministry. It was a subject about which he knew very little; but he rose, as a practised speaker often does, to make a few remarks in a conversational tone, to detect some flaw in a preceding speaker's argument, or to throw light upon some particular section of the subject, without intending to make an elaborate speech, or to review the entire question. Immediately from the lobbies and the coffee-room the members came crowding in, anxious to hear a speaker of whom such great expectations were entertained. He seems to have thought that it would be disrespectful to those members to sit down at once, so he continued extempore, and soon showed his little knowledge of the subject. When he concluded, there was a universal feeling of disappointment. A member named Courtenay rose, and completed his discomfiture by a most virulent and satirical attack, which the rules of the House prevented him from answering. It is hardly necessary to say that Courtenay was an Irishman. He confessed afterwards to Lord Byron that he had been actuated by a personal motive.[1]

After this failure, Flood scarcely ever spoke again. Once, however, in 1790, his genius shone out with something of its old brilliancy in bringing forward a Reform Bill. His proposition was that 100 members, chosen by county household suffrage, should be added to the House; and the speech in which he defended it was much admired by all parties. Burke said that he had retrieved his reputation. Fox declared that his proposition was the best that had been proposed, and Pitt based his opposition to it almost exclusively upon the disturbed state of public affairs. It is to be hoped that these praises in some degree soothed his mind, which must have been bitterly mortified by his previous disappointment. In his reply upon this question, when answering some charge that had been brought against him, he alluded in a very touching manner to the isolation of his position. 'I appeal to you,' he said, 'whether my conduct has been that of an advocate or an agitator; whether I have often trespassed upon your attention; whether ever, except on a question of importance; and whether I then wearied you with ostentation or prolixity. I am as independent

---

[1] Wraxall, speaking of Flood's failure, says: 'The slow, measured, and sententious style of enunciation which characterised his eloquence, however calculated to excite admiration it might be in the senate of the sister kingdom, appeared to English ears cold, stiff, and deficient in some of the best recommendations to attention.' This passage is very curious, as showing how little the present popular conception of Irish eloquence prevailed in the last century.

in fortune and nature as the honourable member himself. I have no fear but that of doing wrong, nor have I a hope on the subject but that of doing some service before I die. The accident of my situation has not made me a partisan; and I never lamented that situation till now that I find myself as unprotected as I fear the people of England will be on this occasion.' After this he only made one other speech —on the French treaty—of any importance. He is said in his last years to have retired much from society, and his temper became gloomy and morose. He died in 1791.

When he felt death approaching he requested his attendant to leave the room, and he drew his last breath alone. Faithful to the end to the interests of his country, he left a large property to the Dublin University, chiefly for the encouragement of the study of Irish, and for the purchase of Irish manuscripts.

There is something inexpressibly melancholy in the life of this man. From his earliest youth his ambition seems to have been to identify himself with the freedom of his country. But though he attained to a position which, before him, had been unknown in Ireland; though the unanimous verdict of his contemporaries pronounced him to be one of the greatest intellects that ever adorned the Irish Parliament; and, though there is not a single act of his life which may not be construed in a sense perfectly in harmony with honour and with patriotism, yet his career presents one long series of disappointments and reverses. At an age when most statesmen are in the zenith of their influence he sank into political impotence. The party he had formed discarded him as its leader. The reputation he so dearly prized was clouded and assailed; the principles he had sown germinated and fructified indeed, but others reaped their fruit, and he is now scarcely remembered except as the object of a powerful invective in Ireland, and as an example of a deplorable failure in England. A few pages of oratory, which probably at best only represent the substance of his speeches, a few youthful poems, a few laboured letters, and a biography so meagre and so unsatisfactory that it scarcely gives us any insight into his character, are all that remain of Henry Flood. The period in which he lived, a jealous and uncertain temper, and two or three lamentable mistakes of judgment, were fatal to his reputation; and he laboured for a people who have usually been peculiarly indifferent to the reputation of their great men. We may say of him as Grattan said of Kirwan: 'The curse of Swift was upon him, to have been born an Irishman and a man of genius, and to have used his talents for his country's good.'

HENRY GRATTAN.

A paper was found in Swift's desk after his death, containing a list of his friends, classified as grateful, ungrateful, and indifferent. In this list the name of Grattan occurs three times, and each time it is marked as grateful. The family was one of some weight in the country, and the father of the subject of the present sketch was Recorder and Member for Dublin. As I have already had occasion to observe, Dr. Lucas was his colleague and his opponent, and a bitter animosity, both personal and political, subsisted between them. The Recorder seems to have been a man of a violent and overbearing temper, firmly wedded to his own opinions, and exceedingly intolerant of contradiction. He was greatly exasperated with his son for adopting Liberal politics, and he carried his resentment so far as to mark his displeasure in his will. Henry Grattan was born in the year 1746. From his earliest youth he manifested the activity of his intellect, and the force and energy of his character. Some foolish nursery tales having produced in his mind those superstitious fears that are so common among children, he determined, when a mere boy, to emancipate himself from their control, and was accustomed to go at midnight into a churchyard near his father's house, where he remained till every qualm of terror had subsided. At the University he distinguished himself greatly, and acquired a passion for the classics, and especially for the great orators of antiquity, that never deserted him through life. Long before he obtained a seat in Parliament he had begun to cultivate eloquence. His especial models were Bolingbroke and Junius, and his method was constant recitation. He learnt by heart certain passages of his speeches, and continually revolved them in his mind till he had eliminated all those almost imperceptible prolixities that exist in nearly every written composition. By this method he brought his sentences to a degree of nervousness and of condensation that is scarcely paralleled in oratory. Several anecdotes are told of the difficulties into which his passion for recitation brought him. On one occasion his landlady in England requested his friends to remove that mad young gentleman who was always talking to himself, or addressing an imaginary person called Mr. Speaker. On another, when apostrophising a gibbet in Windsor Forest, he was interrupted by a tap on the shoulder, and a curious enquiry as to how he had got down. His letters written at this time show that he was subject to violent fits of despondency, and they betray also a morbidness that is singularly unlike his character in after-years.

Shortly after leaving the University he was called to the Bar, and resided for some time in the Temple, where he probably occupied himself much more in the study of oratory than of law. He had obtained access to the House of Lords, and had come completely under the spell of Lord Chatham's eloquence. He wrote an elaborate character of Chatham, which was inserted in 'Baratariana;' and in a letter written some years later he gives a long and very minute description of his style of speaking. The following extract will be read with pleasure, as forming a very vivid description of the most effective of British orators: 'He was very great, but very odd; he spoke in a style of conversation; not, however, what I expected. It was not a speech, for he never came with a prepared harangue. His style was not regular oratory, like Cicero or Demosthenes, but it was very fine and very elevated, and above the ordinary subjects of discourse. . . . Lord Mansfield, perhaps, would have argued the case better; Charles Townshend would have made a better speech; but there was in Lord Chatham a grandeur and a manner which neither had, and which was peculiar to him. What Cicero says in his "Claris Oratoribus" exactly applies: "Formæ dignitas, corporis motus, plenus et artis et venustatis, vocis et suavitas et magnitudo." His gesture was always graceful. He was an incomparable actor; had it not been so he would have appeared ridiculous. His address to the tapestry and to Lord Effingham's memory required an incomparable actor, and he was that actor. His tones were remarkably pleasing. I recollect his pronouncing one word —effete—in a soft, charming accent. His son could not have pronounced it better. He was often called to order. On one occasion he said, "I hope some dreadful calamity will befall the country that will open the eyes of the King;" and then he introduced the allusion to the figure drawing the curtains of Priam, and gave the quotation. He was called to order. He stopped and said, "What I have spoken I have spoken conditionally, but now I retract the condition. I speak absolutely, and I do hope that some signal calamity will befall the country;" and he repeated what he had said. He then fired and oratorised, and grew extremely eloquent. Ministers, seeing what a difficult character they had to deal with, thought it best to let him proceed. On one occasion, addressing Lord Mansfield, he said, "Who are the evil advisers of his Majesty? Is it you? is it you? is it you?" (pointing to the Ministers, until he came near Lord Mansfield). There were several lords round him, and Lord Chatham said, "My Lords, please to take your seats." When they had sat down, he pointed to Lord Mansfield and said, "Is it you? Methinks Felix trembles." It required a great actor to do this. Done by any one else it would have been miserable. . . . . When he came to the argumentative part of his speech, he lowered his tone so as to be scarcely audible; and he did not lay so much stress on those parts as on the great bursts of genius and the sublime passages. He had studied action, and his gesture was graceful, and had a most powerful effect. His speeches required good acting, and he gave it to them. Their impression was great. His manner was dramatic. In this it was said that he was too much of a mountebank, but, if so, it was a great mountebank. Perhaps he was not so good a debater as his son, but he was a much better orator, a better scholar, and a far greater man. Great subjects, great empires, great characters, effulgent ideas, and classical illustrations, formed the material of his speeches.'

It is curious that Grattan, who was so sensible to the advantages of a graceful delivery in others, should have been always remarkable for the extreme singularity and awkwardness of his own. Byron, who otherwise admired his speaking exceedingly, called it a 'harlequin manner.'[1] O'Connell said that he nearly swept the ground with his gestures, and the motion of his arms has been compared to the rolling of a ship in a heavy swell.

While the genius of Chatham had stimulated the ambition of Grattan to the highest degree, the friendship of Flood was directing his enthusiasm in the

---

[1] This was in prose. In his poetry he described Grattan as

'With all that Demosthenes wanted endowed,
And his rival or victor in all he possessed.'
The Irish Avatar.

channel of Irish politics. These two men, afterwards such bitter rivals, were at first intimate friends; and the experience and the counsel of Flood had undoubtedly great influence in moulding the character of Grattan. They declaimed together, they acted together in private theatricals, they wrote together in 'Baratariana,' and they discussed together the prospects of their party.

In 1775 Lord Charlemont brought Grattan into Parliament. The circumstances were, in some respects, very favourable for the display of his genius, for the patriotic party had lost its leader, and there was no one to assert its principles with effect. Grattan cannot with any justice be accused of having supplanted Flood. He simply occupied the position which was vacant, and which his extraordinary eloquence naturally gave him. Whatever opinion might be entertained among his hearers of the wisdom of his political views, or of his judgment, there could be no question that he was from the very commencement of his career by far the greatest orator of the day. When, therefore, the party found themselves deserted by their old leader, they naturally rallied around the one man whose abilities were sufficient to supply his place.

The eloquence of Grattan, in his best days, was in some respects perhaps the finest that has been heard in either country since the time of Chatham. Considered simply as a debater, he was certainly inferior to both Fox and Pitt, and perhaps to Sheridan; but he combined two of the very highest qualities of a great orator to a degree that was almost unexampled. No British orator except Chatham had an equal power of firing an educated audience with an intense enthusiasm, or of animating and inspiring a nation. No

British orator except Burke had an equal power of sowing his speeches with profound aphorisms and associating transient questions with eternal truths. His thoughts naturally crystallised into epigrams; his arguments were condensed with such admirable force and clearness that they assumed almost the appearance of axioms; and they were often interspersed with sentences of concentrated poetic beauty, which flashed upon the audience with all the force of sudden inspiration, and which were long remembered and repeated. Some of his best speeches combined much of the value of philosophical dissertations with all the charm of the most brilliant declamation. I know, indeed, none in modern times, except those of Burke, from which the student of politics can derive so many profound and valuable maxims of political wisdom, and none which are more useful to those who seek to master that art of condensed energy of expression in which he almost equalled Tacitus. His eloquence had nothing of the harmonious and unembarrassed flow of Pitt or of Plunket; and he had no advantages of person and no grace and dignity of gesture; but his strange, writhing contortions, and the great apparent effort he often displayed, added an effect of surprise to the sudden gleams of luminous argument—to the severe and concentrated declamation—to the terseness of statement and the exquisite felicities of expression with which he adorned every discussion. O'Connell, comparing him to Pitt, said that he wanted the sustained dignity of that speaker, but that Pitt's speeches were always speedily forgotten, while Grattan was constantly saying things that were remembered. His speeches show no wit and no skill in the lighter forms of sarcasm; but he was almost unrivalled in crushing invective, in delineations of character, and in brief, keen arguments. In carrying on a train of sustained reasoning he was not so happy. Flood is said to have been his superior; and none of his speeches in this respect are comparable to that of Fox on the Westminster scrutiny.

The extraordinary excellence of his speaking consisted much more in its wonderful positive merits than in its purity or freedom from defects. There was no conscious affectation in his nature, but he had an intense mannerism, which appeared equally in his speaking and in his private life—in almost everything he said or wrote. He rarely said simple things in a simple way; and the quaint peculiarities of his diction appeared as strongly in his conversation and in his unstudied replies as in his elaborate orations. His compositions were almost always overloaded with epigram and antithesis, and his metaphors were often forced, sometimes confused and inaccurate, and occasionally even absurd. But with all these defects very few speakers of any age or country have equalled him in originality, in fire, and in persuasive force. In one respect he would probably have had more influence in our day than in his own, for the reporter's pen would have concealed most of his defects and magnified most of his merits. The political orator now speaks less to those who are assembled within the walls of Parliament than to the public outside. The charm of manner, the music of the modulated tone, have lost their old supremacy, while the power of condensed and vivid expression has acquired an increased value. He who can furnish the watchwords of party, the epigrams of debate, will now exercise the greatest and most abiding influence. A hundred pens will reproduce his words, and they will be repeated as proverbs when the most brilliant displays of diffusive rhetoric are forgotten.

Much of the great influence of the speaking of

Grattan was undoubtedly due to moral causes. There was a certain transparent simplicity and rectitude of purpose, a manifest disinterestedness, a fervid enthusiasm of patriotism in his character, which added greatly to the effect of his eloquence, and gave him an ascendency that was exercised by none of his contemporaries in Ireland. In purely intellectual endowments he was probably equalled by Plunket; but Plunket never exercised even a perceptible influence upon public opinion, while Grattan in a great degree formed the character of the nation. From the very beginning of his career his eloquence became the great vivifying principle in the patriotic party, and every question received a new impulse from his advocacy.

I have already enumerated the principal objects of the party with which Grattan was connected. He assisted Burgh and Flood in carrying the free-trade question to a triumphant issue. He endeavoured, though unsuccessfully, to place the Irish army under the control of the Parliament; and, above all, he gave an unprecedented impulse to the great cause of parliamentary independence. In April 1780 he moved 'that no person on earth, save the King, Lords, and Commons of Ireland, has a right to make laws for Ireland.' This motion he introduced with a speech of splendid eloquence, and the effect produced by it was very great. Flood, however, perceived that it was somewhat premature and would have been defeated, and at his suggestion it was withdrawn. This debate had a considerable effect in eliciting the feelings of the people, and the sentiments of the Parliament are sufficiently shown by the letters of Lord Buckingham, who was then Viceroy, to the Government in England. 'It is with the utmost concern,' he wrote, 'I must acquaint your lordship that although so many gentlemen expressed their concern that the subject had been introduced, the sense of the House against the obligation of any statutes of the Parliament of Great Britain within this kingdom is represented to me to have been almost unanimous.' Shortly after this debate the Volunteer Convention assembled at Dungannon to throw their influence into the scale. Grattan, in co-operation with Flood and Charlemont, drew up a series of resolutions, which were adopted unanimously, asserting the Irish independence; and Grattan, alone, drew up another resolution expressing the gratification with which the Volunteers had witnessed the relaxation of the penal code. It is scarcely possible to exaggerate the importance of this last resolution. It marked the solemn union between the two great sections of Irishmen for the purpose of obtaining the recognition of their country's rights. It showed that the old policy of governing Ireland by the division of her sects had failed; and that if the independence of Parliament were to be withheld, it must be withheld in opposition to a nation united and in arms.

The Government at length yielded. The Duke of Portland was sent over as Lord-Lieutenant, with permission to concede the required boon. At the last moment an effort was made to procure a delay, but Grattan refused to grant it; and on the 16th of April 1782, amid an outburst of almost unparalleled enthusiasm, the declaration of independence was brought forward. On that day a large body of the Volunteers were drawn up in front of the old Parliament House of Ireland. Far as the eye could stretch the morning sun glanced upon their weapons and upon their flags; and it was through their parted ranks that Grattan passed to move the emancipation of his country. Never had a great orator a nobler or a more pleasing task. It was to proclaim that the strife of six centuries had terminated; that the cause for which so much blood had been shed, and so much genius expended in vain, had at last triumphed; and that a new era had dawned upon Ireland. Doubtless on that day many minds reverted to the long night of oppression and crime through which Ireland had struggled towards that conception which had been as the pillar of fire on her path. But now at last the promised land seemed reached. The dream of Swift and of Molyneux was realised. The blessings of independence were reconciled with the blessings of connection; and in an emancipated Parliament the patriot saw the guarantee of the future prosperity of his country and the Shekinah of liberty in the land. It was impossible indeed not to perceive that there was still much to be done—disqualifications to be removed, anomalies to be rectified, corruption to be overcome; but Grattan at least firmly believed that Ireland possessed the vital force necessary for all this, that the progress of a healthy public opinion would regenerate and reform the Irish Parliament as it regenerated and reformed the Parliament of England; and that every year the sense of independence would quicken the sympathy between the people and their representatives. It was indeed a noble triumph, and the orator was worthy of the cause. In a few glowing sentences he painted the dreary struggle that had passed, the magnitude of the victory that had been achieved, and the grandeur of the prospects that were unfolding. 'I am now,' he exclaimed, 'to address a free people. Ages have passed away, and this is the first moment in which you could be distinguished by that appellation. I have spoken on the subject of your liberty so often that I have nothing to add, and have only to admire by what heaven-directed steps you have proceeded until the whole faculty of the nation is braced up to the act of her own deliverance. I found Ireland on her knees; I watched over her with a paternal solicitude; I have traced her progress from injuries to arms, and from arms to liberty. Spirit of Swift, spirit of Molyneux, your genius has prevailed! Ireland is now a nation. In that character I hail her, and, bowing in her august presence, I say *esto perpetua!*'

The concession was made, on the whole, with no ungrudging hand, and in a few years most of the badges of subserviency which the Irish Protestants had worn were discarded. Between 1778 and 1782 the commercial restrictions were almost all abolished; the judges were made immovable; the duration of Parliament was limited; the army was placed in subordination to the Parliament; the appellate jurisdiction of the House of Lords, which had been destroyed in 1719, was restored, and the independence of the Irish Legislature was recognised. Immediately after the concession of independence a day of thanksgiving was appointed to consecrate the triumph, and a vote for the support of twenty thousand sailors for the English navy was agreed upon. This last was almost the first measure of the emancipated Parliament. In this, as in every other period of his career, Grattan was anxious to show in the most unequivocal manner the sympathy of Ireland with England, and the compatibility of an ardent love of independence with a devoted attachment to the connection. He said himself, 'I am desirous above all things, next to the liberty of the country, not to accustom the Irish mind to an alien or suspicious habit with regard to Great Britain.'

While the greatest Irishmen in Ireland were thus working out the freedom of their country, the greatest

Irishman in England wrote to encourage them and to express his approval of the work. 'I am convinced,' wrote Burke to Lord Charlemont, 'that no reluctant tie can be a strong one, and that a natural, cheerful alliance, will be a far more secure link of connection than any principle of subordination borne with grudging and discontent.' The Whig party, who were for a brief period in power, appear to have concurred in this view; and Fox, in one of his speeches in 1797, expressed it very unequivocally. 'I would have the Irish government,' he said, 'regulated by Irish notions and Irish prejudices, and I am convinced that the more she is under Irish government the more she will be bound to English interests.'[1]

The Parliament at this time determined to mark its recognition of the services of Grattan by a grant of 100,000l. Grattan, however, refused to receive so large a sum, and was with some difficulty induced to accept half. This grant enabled him to devote himself exclusively to the service of the country without practising at the Bar, to which he had been called.

I need not revert at length to the question of Simple Repeal, which I have already so fully considered. The arguments on each side of that controversy must be admitted to have been very nicely balanced, and the authorities were also very evenly divided. Grattan reckoned among the supporters of his view Charlemont, Fox, the Irish chief justices and chief baron, and several other Irish legal authorities. He had, however, injured his cause greatly by bringing forward a resolution declaring that all who asserted that England had authority over Ireland were enemies to the country—a resolution which was wholly indefensible, which Flood most triumphantly assailed, and which, after a short

[1] Quoted by Lord J. Russell in 1837. See Ann. Reg. 1837, p. 31.

discussion, was withdrawn. He was also, as it appears to me, guilty of a grave error in not urging at this time more vehemently the question of Parliamentary Reform. After their famous conflict, the two rivals co-operated successfully in opposing some commercial arrangements known as Orde's Propositions, which were brought forward in 1785, and which, by denying the Irish Parliament the right of initiation on commercial matters, trenched upon the independence of Ireland.

In December 1783 Pitt's Ministry began. It appears, from one of the letters of Pitt, that at the beginning of his career he contemplated reforming the Irish as well as the English Parliament; but in this, as in nearly every portion of his policy, he speedily apostatised to the views of the Tory party, who had brought him into power, and resisted every remedial measure which was likely to prove in the least embarrassing or dangerous to his Ministry. During many years of the Ministry of Pitt which preceded the Union, the Irish administration almost uniformly opposed every effort to reform the Parliament. One of the greatest causes of complaint was the Pension List. The enormity of the grievance is sufficiently shown by the fact that the money spent in pensions in Ireland was not merely relatively, but absolutely, greater than was expended for that purpose in England; that the pension list trebled in the first thirty years of George III.; and that in 1793 it amounted to no less than 124,000*l.* Repeated efforts were made to reduce this list, which was so detrimental to the disordered finances of the country, and so fatal to the purity of Parliament. Grattan brought forward the subject in 1785 and in 1791, but on both occasions Government threw their influence into the opposite scale, and he was defeated.

In 1789 Grattan disagreed with Pitt's Ministry on the Regency question, and maintained with Fox that the madness of the King was to be regarded as tantamount to his death, and that while it lasted his son rightfully possessed the full powers of royalty. The Irish Parliament adopted this view, and there was some danger of a serious collision with England, when the recovery of the King solved the difficulty. But the great question which at this time agitated the public mind was the position of the Roman Catholics—a question which has long been the most fertile cause of dissension and controversy in Ireland.

There are few more curious pages in ecclesiastical history than that which records the various phases of Christianity in Ireland. Its first introduction is lost in the obscurity of antiquity, but we find it existing, though in a very feeble condition, in the middle of the fifth century, when Palladius and St. Patrick came over to re-animate it. Palladius was sent from Rome by Pope Celestine; his mission was wholly unsuccessful, and he very soon left Ireland. From what quarter St. Patrick derived his authority is a question which is still fiercely debated between the members of the rival creeds. It seems plain that under his auspices Christianity spread over the entire island; that the Church continued for several centuries in a flourishing condition; that it existed very independently of Rome; and that in the famous Easter controversy it warmly upheld the Oriental opinion.

The Irish monasteries soon became famed for the piety and the learning that emanated from them, and many pilgrims from many lands sought instruction within their walls. Amongst others, Oswald, the son of the King of Northumbria, was educated and converted to Christianity by the Irish monks; and, when he came to the throne, he invited his old preceptors to plant a mission in his dominions, and established the monastery of Lindisfarne. It was the rare fortune of the monks of Lindisfarne to have three successive priors who were so stainless in their character, so winning in their manners, and so gentle in their controversies, that they prepossessed all who knew them in behalf of their religion, and extorted expressions of the warmest admiration even from an historian[1] who was an opponent of their views. Their zeal was equal to their gentleness, and their success to their deserts, and by their means the light of Christianity was spread over nearly the whole of the north of England. At last, however, they came into collision with the Roman party on the Easter question; and the genius and the energy of Wilfrid, the Roman champion, having gained the victory, they returned to their own country. In Ireland the Pope obtained a certain influence amid the civil wars that distracted the land, but his authority was never generally recognised till the English invasion. The English King, having obtained letters from two successive Pontiffs conferring Ireland upon him, on account of its separation from the See of Rome, and on condition of the payment of Peter's pence, convened a council at Cashel, which formally imposed the Roman yoke on the nation from which England had received a Christianity separate from Rome.

If we overleap the next few centuries, we find that at the time of the Reformation Ireland was the only northern country in which the reformed tenets never made way. The explanation of this phenomenon is beyond all question to be found in the policy of England. The Irish regarded Protestantism as identified with a nation which was the object of their deepest

[1] Bede.

abhorrence. Elizabeth, who was its great representative, had spread desolation and disaster over the greater part of their land. She had shown herself anxious to propagate the Reformed faith, but still more anxious to eradicate the nationality of Ireland. To effect the former object she enjoined that the Anglican service should everywhere be celebrated; to effect the latter she forbade its being celebrated in the Irish tongue. Where the people could not understand English, it was gravely ordered that the service might be translated into Latin. The consequence was what might have been anticipated. The people continued in their old faith, and England was thus the means of consolidating and perpetuating that religion which has ever proved the most insuperable obstacle to her policy.

The next great representative of Protestantism in England was Cromwell, whose Irish policy is well known. An illustrious living writer has discovered a transcendent, and even religious, grandeur in the massacres of Drogheda and of Wexford, but it must be admitted that they were not calculated to prepossess the Irish mind in favour of Protestantism. We may observe, too, that the Puritans acted throughout as religionists. Every soldier was an ardent theologian, and never more so than when, with a text from Joshua in his mouth, he was hewing the misbeliever to the ground. The war of races and the recollection of the Irish massacre seem to have all given way to the fierce hatred of the Man of Sin, that had steeled every heart and whetted every sword. Had Cromwell's policy been persisted in for a few generations, Catholicism in Ireland might have perished in blood; but, as it was, it only deepened the chasm between the two religions, and inspired the Roman Catholics with a still more intense hatred of the dominant creed.

The last great Protestant ruler of England was William III., who is identified in Ireland with the humiliation of the Boyne, with the destruction of Irish trade, and with the broken treaty of Limerick. The ceaseless exertions of the extreme Protestant party have made him more odious in the eyes of the people than he deserves to be; for he was personally far more tolerant than the great majority of his contemporaries, and the penal code was chiefly enacted under his successors. It required, indeed, four or five reigns to elaborate a system so ingeniously contrived to demoralise, to degrade, and to impoverish the people of Ireland. By this code the Roman Catholics were absolutely excluded from the Parliament, from the magistracy, from the corporations, from the bench, and from the bar. They could not vote at parliamentary elections or at vestries. They could not act as constables, or sheriffs, or jurymen, or serve in the army or navy, or become solicitors, or even hold the positions of gamekeeper or watchman. Schools were established to bring up their children as Protestants; and if they refused to avail themselves of these, they were deliberately consigned to hopeless ignorance, being excluded from the University, and debarred, under crushing penalties, from acting as schoolmasters, as ushers, or as private tutors, or from sending their children abroad to obtain the instruction they were refused at home. They could not marry Protestants; and if such a marriage were celebrated it was annulled by law, and the priest who officiated might be hung. They could not buy land, or inherit or receive it as a gift from Protestants, or hold life annuities, or leases for more than thirty-one years, or any lease on such terms that the profits of the land exceeded one-third of the rent. If any Catholic leaseholder by his industry so increased his profits that they exceeded this proportion, and did not immediately make a corresponding increase in his payments, any Protestant who gave the information could enter into possession of his farm. If any Catholic had secretly purchased either his old forfeited estate, or any other land, any Protestant who informed against him might become the proprietor. The few Catholic landholders who remained were deprived of the right which all other classes possessed of bequeathing their lands as they pleased. If their sons continued Catholics, it was divided equally between them. If, however, the eldest son consented to apostatise, the estate was settled upon him, the father from that hour became only a life tenant, and lost all power of selling, mortgaging, or otherwise disposing of it. If the wife of a Catholic abandoned the religion of her husband, she was immediately free from his control, and the Chancellor was empowered to assign to her a certain proportion of her husband's property. If any child, however young, professed itself a Protestant, it was at once taken from the father's care, and the Chancellor could oblige the father to declare upon oath the value of his property, both real and personal, and could assign for the present maintenance and future portion of the converted child such proportion of that property as the court might decree. No Catholic could be guardian either to his own children or to those of another person; and therefore a Catholic who died while his children were minors had the bitterness of reflecting upon his death-bed that they must pass into the care of Protestants. An annuity of from twenty to forty pounds was provided as a bribe for every priest who would become a Protestant. To convert a Protestant to Catholicism was a capital offence. In every walk of life the Catholic was pursued by persecution or restriction. Except in the linen trade, he could not have more than two apprentices. He could not possess a horse of the value of more than five pounds, and any Protestant, on giving him five pounds, could take his horse. He was compelled to pay double to the militia. He was forbidden, except under particular conditions, to live in Galway or Limerick. In case of war with a Catholic power, the Catholics were obliged to reimburse the damage done by the enemy's privateers. The Legislature, it is true, did not venture absolutely to suppress their worship, but it existed only by a doubtful connivance,—stigmatised as if it were a species of licensed prostitution, and subject to conditions which, if they had been enforced, would have rendered its continuance impossible. An old law which prohibited it, and another which enjoined attendance at the Anglican worship, remained unrepealed, and might at any time be revived; and the former was, in fact, enforced during the Scotch rebellion of 1715. The parish priests, who alone were allowed to officiate, were compelled to be registered, and were forbidden to keep curates, or to officiate anywhere except in their own parishes. The chapels might not have bells or steeples. No crosses might be publicly erected. Pilgrimages to the holy wells were forbidden. Not only all monks and friars, but also all Catholic archbishops, bishops, deacons, and other dignitaries, were ordered by a certain day to leave the country; and if after that date they were found in Ireland they were liable to be first imprisoned and then banished; and if after that banishment they returned to discharge their duty in their dioceses, they were liable to the punishment of death. To facilitate the discovery of offences against the code, two justices of the peace might at any time compel any Catholic of eighteen years of age to declare when and where he last heard mass, what persons were present, and who officiated; and if he refused to give evidence they might imprison him for twelve months, or until he paid a fine of twenty pounds. Anyone who harboured ecclesiastics from beyond the seas was subject to fines which for the third offence amounted to the confiscation of all his goods. A graduated scale of rewards was offered for the discovery of Catholic bishops, priests, and schoolmasters; and a resolution of the House of Commons pronounced 'the prosecuting and informing against Papists' 'an honourable service to the Government.'

Such were the principal articles of this famous code —a code which Burke truly described as 'well digested and well disposed in all its parts; a machine of wise and elaborate contrivance, and as well fitted for the oppression, impoverishment, and degradation of a people, and the debasement in them of human nature itself, as ever proceeded from the perverted ingenuity of man.' It was framed by a small minority of the nation for the oppression of the majority who remained faithful to the religion of their fathers. It was framed by men who boasted that their creed rested upon private judgment, and whose descendants are never weary of declaiming upon the intolerance of Popery; and it was directed, in many of its provisions, against mere religious observances; and was in all its parts so strictly a code of religious persecution, that any Catholic might be exempted from its operation by simply forsaking his religion. It was framed and enforced, although by the Treaty of Limerick the Catholics had been guaranteed such privileges in the exercise of their religion as they enjoyed in the reign of Charles II., although the Sovereign at the same time promised, as soon as his affairs would permit, 'to summon a Parliament in this kingdom, and to endeavour to procure the said Roman Catholics such further security in that particular as may preserve them from any disturbance on account of their religion,' although not a single overt act of treason was proved against them, and although they remained passive spectators of two rebellions which menaced the very existence of the Protestant dynasty in England.

It is impossible for any Irish Protestant, whose mind is not wholly perverted by religious bigotry, to look back without shame and indignation to the penal code. The annals of persecution contain many more sanguinary pages. They contain no instance of a series of laws more deliberately and ingeniously framed to debase their victims, to bribe them in every stage of their life to abandon their convictions, and to sow dissension and distrust within the family circle. That the Irish Parliament, in the last years of William, and in the reigns of his two successors, was one of the most persecuting legislative assemblies that have ever sat, cannot reasonably be questioned. But, without descending to the moral sophistry which some writers have employed in endeavouring to palliate these laws, there is something that may be truly said for the Irish Protestants. The laws which had been passed in England and by the English Parliament under William, for the oppression of Catholics, were on the whole even more stringent than those which were subsequently passed in Ireland, and some of the worst Irish Acts were simply transcripts of English laws. The beginning of the Irish penal code was a law passed in 1691 by the English Parliament for excluding all Catholics from the Irish one. The Irish Protestants sometimes surpassed in bigotry the wishes of the English Cabinet, but yet a long succession of Lord-Lieutenants, speaking as the representatives of the English Government, urged increased severity against the 'common enemy,' and among these Governors we find such men as Carteret and Chesterfield. The spirit in which Ireland was systematically governed in the early part of the eighteenth century was well illustrated by the speech of the Lords Justices to the Parliament in 1715, in which they said, 'We must recommend to you, in the present conjuncture, such unanimity in your resolutions as may once more put an end to all other distinctions in Ireland than that of Protestant and Papist.' The time when the Irish Parliament was most persecuting, and the Irish Protestants were most fanatical, was the time when the first was absolutely subservient to foreign control, and when the latter considered themselves merely as a garrison in an enemy's country. No sooner had a national spirit arisen among the Protestants than the spirit of sectarianism declined. The penal laws were never for any considerable time enforced in their full severity, and some parts of them —especially those restricting the Catholic worship, banishing bishops and friars, and prohibiting Catholic schools—became in the latter and the greater part of their existence a mere dead letter. Much property that would otherwise have passed to Protestants was retained in Catholic hands by legal fictions and by the assistance or with the connivance of Protestants, and the emancipated Parliament of Ireland carried the policy of religious liberty much farther than the Parliament of England.

The economical and moral effects of the penal laws were, however, profoundly disastrous. The productive energies of the nation were fatally diminished. Almost all Catholics of energy and talent who refused to abandon their faith emigrated to foreign lands. The relation of classes was permanently vitiated; for almost all the proprietary of the country belonged to one religion, while the great majority of their tenants were of another. The Catholics, excluded from almost every possibility of eminence, deprived of their natural leaders, and consigned by the Legislature to utter ignorance, soon sank into the condition of broken and dispirited helots. A total absence of industrial virtues, a cowering and abject deference to authority, a recklessness about the future, a love of secret illegal combinations, became general among them. Above all, they learnt to regard law as merely the expression of force, and its moral weight was utterly destroyed. For the greater part of a century the main object of the Legislature was to extirpate a religion by the encouragement of some of the worst and the punishment of some of the best qualities of our nature. Its rewards were reserved for the informer, for the hypocrite, for the undutiful son, or for the faithless wife. Its penalties were directed against religious constancy and the honest discharge of ecclesiastical duty.

It would, indeed, be scarcely possible to conceive a more infamous system of legal tyranny than that which in the middle of the eighteenth century crushed every class and almost every interest in Ireland. The Parliament had been deprived of every vestige of independence. The English House of Lords, by an act of what appears to have been pure usurpation, had in 1719 assumed to itself the right of final judicature in Irish cases, and deprived the Irish House of Lords of all judicial powers. The English Chancellor, Lord Macclesfield, had laid down the doctrine that 'the English Courts of Justice have a superintendent power over those of Ireland,' and are able to reverse their sentences. The Irish judges might at any time be removed. Manufacturing and commercial industry had been deliberately crushed for the benefit of English manufacturers, and the country was reduced to such a state of poverty that in 1779 the Government was compelled to borrow 50,000*l*. from England and 20,000*l*. from a private individual, to pay its troops. At the same time a gigantic and ever-increasing pension list was drawn from the scanty resources of the nation, and was expended partly in corrupting its representatives and partly in rewarding Englishmen or foreigners. The mistresses of George I., the Queen Dowager of Prussia, sister of George II., the Sardinian ambassador who negotiated the Peace of Paris, were all on the Irish pension lists. The most honourable and most lucrative positions in Ireland were chiefly held by Englishmen. The Lord-Lieutenant, the Chief Secretary, and most of the other foremost political officers, were always Englishmen. During the whole of the eighteenth century there was not a single instance of an Irishman holding the office of Archbishop of Armagh; and of the eighteen Archbishops of Dublin and Cashel, ten were Englishmen, as were also nearly all the chancellors and a large proportion of the bishops and judges. And, while even the favoured minority of the Irish people were thus systematically depressed, the great majority were deprived of all political privileges, excluded from almost all means of acquiring weath, reduced by law into a pariah class, and exposed to demoralising influences which even to the present day have left their traces upon the national character.

There can be no question that in the higher ranks of Catholics the penal laws produced a large amount of formal apostacy. The desire of the Catholic landlord to keep his property in his family was often stronger than his religious feeling, and among professional men very little scruple appears to have been felt. In a remarkable letter to the Duke of Newcastle, written in the early part of 1727, Primate Boulter complains that 'the practice of the law from the top to the bottom is at present mostly in the hands of new converts, who give no farther security on this account than producing a certificate of their having received the Sacrament in the Church of England or Ireland, which several of those who were Papists obtain on the road hither, and demand to be admitted barrister in virtue of it at their arrival; and several of them have Popish wives, and mass said in their houses, and breed up their children Papists. Things are at present so bad with us that if about six should be removed from the the Bar to the Bench here, there will not be a barrister of note left that is not a convert.' In order to check this state of things, a number of enactments were made to compel converts to educate their children as Protestants, and to subject those who refused and those who married Papist wives to the same disabilities as if they had not professed themselves Protestants. But the movement of conversion to Protestantism was only in the upper classes; and it is a singularly curious fact that at the worst period of the penal laws poor Protestants were continually lapsing into Catholicism, while the poor Catholics remained steadfast in their faith. Primate Boulter, to check the movement, founded the charter schools, which were intended to be the only means of educating the Irish poor, and which were essentially proselytising. 'I can assure you,' he writes to the Bishop of London, 'the Papists are here so numerous that it highly concerns us in point of interest, as well as out of concern for the salvation of those poor creatures who are our fellow-subjects, to try all possible means to bring them and theirs over to the knowledge of the true religion; and one of the most likely methods we can think of is, if possible, instructing and converting the young generation; for instead of converting those that are adult, we are daily losing several of our meaner people, who go off to Popery. .... The ignorance and obstinacy of the adult Papists is such that there is not much hope of converting them.'

The history of the penal laws should, indeed, furnish a lasting warning to persecutors of all religions. Arthur Young asserts that the numerical proportion of the Roman Catholics in Ireland was not even diminished, if anything the reverse; and that it was admitted, by those who asserted the contrary, that it would take 4,000 years, according to the then rate of progress, to convert them. It was stated in Parliament that only 4,055 had conformed in 71 years under the system; and what little the religion may have lost in number it gained in intensity. The poorer classes in Ireland emerged from their long ordeal, penetrated with an attachment to their religion almost unparalleled in Europe. With the exception of the inhabitants of Bavaria and the Tyrol, there is, perhaps, no nation in Europe whose character has been so completely moulded and permeated by it, or in which sceptical doubts are more completely unknown.

The code perished at last by its own atrocity. It became after a time so out of harmony with the prevailing tone of Irish opinion that it ceased to be enforced, and the Irish Protestants took the initiative in obtaining its mitigation. In 1768 a Bill for this purpose passed without a division in the Irish Parliament, but was lost in England. In 1774, 1778, 1782, and 1792, several Relief Bills became law. By these Acts the Roman Catholics were admitted to most of the privileges of their fellow-subjects, except to political power. They still laboured under three great disqualifications: they could not possess the elective franchise, they could not sit in Parliament, and they could not rise to the higher positions in the legal or the military professions. Public opinion had begun to show itself in their favour.[1] As I have already noticed, the Volunteers, who were at first exclusively Protestants, and who were recruited chiefly in the North, soon admitted Catholics into their ranks, and would probably have gone further but for the influence of Charlemont and Flood. Burke espoused their cause warmly, wrote a petition for them, exerted all his eloquence in their behalf, and sent over his son to assist them. But the man to whom they owed the most was undoubtedly Henry Grattan. He was almost the only Irishman of note in Ireland who at that time ceaselessly advocated their unqualified

---

[1] An acute observer, writing in 1770, thus described the religious state of Ireland: 'The rigour of Popish bigotry is softening very fast, the Protestants are losing all bitter remembrance of those evils which their ancestors suffered, and the two sects are insensibly gliding into the same common interests. The Protestants, through apprehensions from the superior numbers of the Catholics, were eager to secure themselves in the powerful protection of an English Minister, and to gain this were ready to comply with his most exorbitant demands; the Catholics were alike willing to embarrass the Protestants as their natural foes; but awakening from this delusion, they begin to condemn their past follies, reflect with shame on having so long played the game of an artful enemy, and are convinced that without unanimity they never can obtain such consideration as may entitle them to demand, with any prospect of success, the just and common rights of mankind. Religious bigotry is losing its force everywhere. Commercial and not religious interests are the objects of almost every nation in Europe.'
—*Preface to the edition of Molyneux's ' Case of Ireland,' which appeared in* 1770.

emancipation. Flood, Charlemont, and Lucas had a different theory. They foresaw that the admission of the Roman Catholics to political equality would sooner or later prove incompatible with the establishment of the Church of the minority; they were not prepared to surrender that establishment, and they therefore maintained that while the Roman Catholics should be admitted to perfect toleration, they should not be admitted to political power. This distinction Grattan refused to recognise. He argued that to exclude the great bulk of the people from Parliament on account of their religion was to inflict upon them a positive injury, and to deprive them of all security for their toleration. 'Civil and religious liberty,' he said in one of his speeches, 'depends on political power; the community that has no share directly or indirectly in political power has no security for its political liberty.' He supported the establishment warmly and consistently,[1] but he made a vigorous effort, in 1788, to substitute some other mode of payment for the tithes, which were chiefly taken from the Roman Catholics. He believed also, like most eminent men of his generation, that the difference between the two religions was much exaggerated; that it was continually lessening, and that the process of assimilation would be greatly accelerated by the removal of the religious disabilities. His speeches are full of intimations of this opinion. 'Bigotry may survive persecution, but it can never

[1] In a letter to the Lord Mayor and Sheriffs of Dublin, on June 1, 1792, he said: 'I love the Roman Catholic; I am a friend to his liberty, but it is only inasmuch as his liberty is entirely consistent with your ascendency, and an addition to the strength and freedom of the Protestant community.'—*Miscellaneous Works*, p. 282. An Irish Protestant in the last century could perhaps hardly write or think otherwise, and it by no means follows that if he lived now his opinion would be the same.

survive toleration.' 'What Luther did for us, philosophy has done in some degree for the Roman Catholics, and their religion has undergone a silent reformation; and both divisions of Christianity, unless they have lost their understanding, must have lost their animosity, though they have retained their distinctions.' 'It is the error of sects to value themselves more upon their differences than upon their religion.'

Among the Roman Catholics themselves, for a considerable time, scarcely any political life had existed. About the middle of the century, it is true, three Catholic writers, named O'Connor, Wyse, and Curry, made laudable efforts to arouse them; but their spirits were completely cowed by long oppression, and the restrictions on education had prevented the development of their intellect. At last, however, Father O'Leary, a writer of real genius, rose among them. It is impossible to read his works without regretting that an eloquence of such extraordinary brilliancy was not exerted more frequently, and on works of greater magnitude. His principal performances are a series of masterly letters to Wesley, who had written against the removal of the penal laws; an address to the Roman Catholics, inculcating loyalty during the Rebellion of 1745; and a short treatise on the Socinian controversy. In England he is scarcely known except by his happy retort to a Protestant bishop, to whose picture of the horrors of purgatory he replied, 'Your lordship might go farther and fare worse;' but his name is still popular in Ireland, and his writings are well worthy of perusal, if it were only for the great beauty of their style. He was in his day beyond all comparison the most brilliant writer in Ireland; and had he moved in a wider sphere, and written on subjects of more enduring value, he might have taken a place among the great masters of English prose. He was admitted a member of a convivial society called the 'Monks of the Screw,' which was presided over by Curran, and which included all the first men in the country. It is a slight but significant fact, that when on one occasion he went to the Volunteer Convention, the Volunteer guard turned out and presented arms to this Catholic priest. He attained a position in Ireland which no member of his order had held for more than a century; his writings were widely read, and Grattan panegyrised him in Parliament. The concluding sentence of that panegyric is curiously characteristic of the speaker, of his subject, and of the theological temperature of their time. 'If I did not know him,' he said, 'to be a Christian clergyman, I should suppose him by his writings to be a philosopher of the Augustan age.'

With this exception, the Catholics seem to have made scarcely any exertion to improve their condition until 1792 and 1793, when they formed a convention, under a leader named Keogh, for the purpose of preparing petitions to the King and to the Parliament.

Grattan conducted their cause with great tact. He refused to make it a party question, and by this refusal obtained the assistance of Sir Hercules Langrishe, who was one of the ablest of his political opponents, and left it always open to the Ministers to adopt his views. At last, in 1793, a Relief Bill, admitting the Roman Catholics to the elective franchise, was introduced by the Government, and, after a warm debate, was carried. In the course of the discussion Grattan made the following statement of the case: 'The situation of the Roman Catholics is reducible to four propositions. They are three-fourths of your people paying their proportion of near 2,000,000*l.* of taxes, without any share in the representation or expenditure; they pay your Church establishment without any retribution; they discharge the active and laborious offices of life, manufacture, husbandry, and commerce, without those franchises which are annexed to the fruits of industry; and they replenish your armies and navies without commission, rank, or reward. Under these circumstances, and under the further recommendation of total and entire political separation from any foreign prince or pretender, they desire to be admitted to the franchise of the constitution.' While supporting the Government Act, Grattan complained greatly of its imperfection. The admission of the Roman Catholics to Parliament was its necessary complement, and by one bold measure the Ministers might have set the question at rest for ever. The measure of 1793 conferred political power on the uneducated masses, while it retained the disqualification of the educated few. Had emancipation at this time been conceded, the great Catholic landlords, being brought forward prominently in the parliamentary arena, would have become the natural leaders of their co-religionists, and the Irish Catholic landlords have always been as loyal, as moderate, and as enlightened as the Protestant ones. But the penal laws having reduced this class to the smallest dimensions, and Tory obstinacy having deprived them of the means of acquiring their legitimate political influence, it is not surprising that the formation of Catholic opinion should have ultimately devolved upon agitators and priests. A Bill for completing the relief was at this time actually brought forward, but was defeated by Government influence.

The Relief Bill of '93 naturally suggests a consideration of the question so often agitated in Ireland, whether the Union was really a benefit to the Roman Catholic cause. It has been argued that Catholic emancipation was an impossibility as long as the Irish Parliament lasted; for in a country where the great majority were Roman Catholics, it would be folly to expect the members of the dominant creed to surrender a monopoly on which their ascendency depended. The arguments against this view are, I believe, overwhelming. The injustice of the disqualification was far more striking before the Union than after it. In the one case the Roman Catholics were excluded from the Parliament of a nation of which they were the great majority; in the other they were excluded from the Parliament of an empire in which they were a small minority. Grattan, Plunket, Curran, Burrowes, and Ponsonby were the great supporters of Catholic emancipation, and the great opponents of the Union. Clare and Duigenan were the two great opponents of emancipation, and the great supporters of the Union. At a time when scarcely any public opinion existed in Ireland, when the Roman Catholics were nearly quiescent, and when the leaning of Government was generally illiberal, the Irish Protestants admitted their fellow-subjects to the magistracy, to the jury-box, and to the franchise. By this last measure they gave them an amount of political power which necessarily implied complete emancipation. Even if no leader of genius had risen in the Roman Catholic ranks, and if no spirit of enthusiasm had animated their councils, the influence possessed by a body who formed three-fourths of the population, who were rapidly rising in wealth, and who could send their representatives to Parliament, would have been sufficient to ensure their triumph.[1] If the Irish Legislature had

---

[1] This was the opinion expressed by Fox in one of his letters soon after the recall of Lord Fitzwilliam : 'As to the Catholic Bill, it is not only right in principle, but, after all that was given to the Catholics

continued, it would have been found impossible to resist the demand for reform; and every reform, by diminishing the overgrown power of a few Protestant landholders, would have increased that of the Roman Catholics. The concession accorded in 1793 was, in fact, far greater and more important than that accorded in 1829, and it placed the Roman Catholics, in a great measure, above the mercy of Protestants. But this was not all. The sympathies of the Protestants were being rapidly enlisted in their behalf. The generation to which Charlemont and Flood belonged had passed away, and all the leading intellects of the country, almost all the Opposition, and several conspicuous members of the Government, were warmly in favour of emancipation. The rancour which at present exists between the members of the two creeds appears then to have been almost unknown, and the real obstacle to emancipation was not the feelings of the people,[1] but the policy of the Government. The Bar may be considered on most subjects a very fair exponent of the educated opinion of the nation; and Wolfe Tone observed, in 1792, that it was almost unanimous in favour of the Catholics; and it is not without importance, as showing the tendencies of the rising generation, that a

two years ago, it seems little short of madness (and at such a time as this) to dispute about the very little that remains to be given them. To suppose it possible that now they are electors they will long submit to be ineligible, appears to me to be absurd beyond measure; but common sense seems to be totally lost out of the councils of this devoted country.'
—*Lord Russell's Life of Fox*, vol. iii. p. 73.

[1] The testimony of Lord Sheffield (who was adverse to the proposition of giving votes to the Catholics) to the feelings of the Irish Protestants is very remarkable. He says, in a work published in 1785: 'The right of being elected would surely follow their being eligible; but, at all events, the power would be in the electors. It is curious to observe one-fifth or perhaps one-sixth of a nation in possession of the power and property of the country, eager to communicate that power to the remaining four-fifths, which would in effect entirely transfer it from themselves.'—*Observations on the Trade of Ireland*, p. 372.

large body of the students of Dublin University in 1795 presented an address to Grattan, thanking him for his labours in the cause. The Roman Catholics were rapidly gaining the public opinion of Ireland, when the Union arrayed against them another public opinion which was deeply prejudiced against their faith, and almost entirely removed from their influence. Compare the twenty years before the Union with the twenty years that followed it, and the change is sufficiently manifest. There can scarcely be a question that if Lord Fitzwilliam had remained in office the Irish Parliament would readily have given emancipation. In the United Parliament for many years it was obstinately rejected, and if O'Connell had never arisen it would probably never have been granted unqualified by the veto. In 1828, when the question was brought forward in Parliament, 61 out of 93 Irish members, 45 out of 61 Irish county members, voted in its favour. Year after year Grattan and Plunket brought forward the case of their fellow-countrymen with an eloquence and a perseverance worthy of their great cause; but year after year they were defeated. It was not till the great tribune had arisen, till he had moulded his co-religionists into one compact and threatening mass, and had brought the country to the verge of revolution, that the tardy boon was conceded. Eloquence and argument proved alike unavailing when unaccompanied by menace, and Catholic emancipation was confessedly granted because to withhold it would be to produce a rebellion.

The refusal of the Government to complete the enfranchisement of the Roman Catholics had a great influence in stimulating disloyalty in the country, but most especially among the Protestants. The conviction that the removal of all religious disabilities was essential to the welfare and to the security of the independence of Ireland, was rapidly gaining ground. In 1782, as we have already seen, the representatives of 143 corps of Volunteers passed a resolution, with but two dissentient voices, expressing their approval of the mitigation of the penal code. In 1792 a petition for emancipation, signed by 600 Protestant householders of Belfast, was presented to the Parliament. In 1791 the club of United Irishmen had been formed, to advocate the Catholic claims. This club consisted originally chiefly of Protestants, who were under no obligation to secrecy, and who were merely pledged 'to promote a union of friendship between Irishmen of every religious persuasion, and to forward a full, fair, and adequate representation of all the people in Parliament.' It was presided over by Hamilton Rowan, a Protestant gentleman of large fortune, and a most amiable and chivalrous character; and it was at first of a perfectly loyal character. Grattan was not in any way connected with it; but, like all the Liberals of the time, he was labouring for the attainment of its two great objects — Catholic emancipation and Parliamentary reform. The latter subject he brought forward, in conjunction with Ponsonby, in 1793. He stated in his speech that less than ninety individuals returned a majority of the Parliament; but he was unable to pass his Bill. There appears indeed to be little question that during the later years of the Ministry of Pitt it was the firm resolution of the Government not only to resist the attempts to purify the Parliament, but also steadily and deliberately to increase its corruption. Fitzgibbon, afterwards Lord Clare, was the chief agent in attaining this end. His avowed political maxim was that 'the only security for national concurrence is a permanent and commanding influence of the English Executive, or rather English Cabinet, in the councils of Ireland;' and for many years before the

Union the Government was continually multiplying places, in order to increase that influence. Grattan described the country as placed 'in a sort of interval between the cessation of a system of oppression and the formation of a system of corruption;' and he scarcely exaggerated the proceedings of the Irish rulers when he described them as 'a set of men possessing themselves of civil, military, and ecclesiastical authority, and using it with a fixed and malignant intention to corrupt the morals, in order to undermine the freedom, of the people.' In 1787 a Peace Preservation Bill cancelled the whole magistracy; and, in addition to many other appointments, gave a salaried and judicial position to thirty-two barristers. In 1789 no less than sixteen peers were created or promoted, and the pension list was increased by 13,000*l.* a year. Ponsonby, at this time, declared that there were 110 placemen in the House of Commons, and that one-eighth of the revenue of the country was divided between members of Parliament. Five more Treasury places were created in 1793. The disfranchisement of revenue officers had been carried in England with general approval and with excellent effect; but the repeated efforts of Grattan to carry it in Ireland were invariably defeated by the Government, and the utmost that the patriots could procure was the permanent reduction of the pension list to 80,000*l.* a year, a Bill by which those who accepted office were unable to sit in Parliament without re-election, and a little more direct Parliamentary control over the pension list.[1]

Under these circumstances, it is not surprising that the many elements of disloyalty and of turbulence which smouldered in the country should have acquired

[1] I have taken most of these facts from Grattan's Life, by his son, which is probably the best history of Ireland at the period under consideration.

new strength. Whoever desires to understand the manner in which they were developed should study the clear and evidently truthful memoir on the rise and aims of the United Irishmen, which was drawn up by their three leaders, O'Connor, Emmett, and Macnevin, when State prisoners.[1] The society, they tell us, was at first simply and frankly loyal, aiming solely at Parliamentary reform and Catholic emancipation, and valuing the latter chiefly as a condition or an element of the former. But, even in 1791, 'it was clearly perceived that the chief support of the borough influence in Ireland was the weight of English influence.' About 1795 the persistent and successful opposition of the Government to reform made the United Irishmen for the first time disloyal. 'They began to be convinced that it would be as easy to obtain a revolution as a reform, so obstinately was the latter resisted; and, as this conviction impressed itself on their minds, they were inclined, not to give up the struggle, but to extend their views. . . . Still,' they add, 'the whole body, we are convinced, would have rejoiced to stop short at reform.' They tried to avail themselves of French assistance, because 'they perceived that their strength was not, and was not likely to become, equal to wresting from the English and the borough interests in Ireland even a reform.' They decided ultimately upon making separation rather than reform their ideal, because 'foreign assistance could only be hoped for in proportion as the object to which it would be applied was important to the party giving it. A reform in the Irish Parliament was no object to the French; a separation of Ireland from England was a mighty one indeed.'

In addition to these considerations, we must remember that the moral influence of the French Revo-

[1] 'Castlereagh Correspondence,' vol. i.

lution had begun to operate upon the country. It is difficult for us, among whom the principles it enunciated are now regarded as mere truisms, to realise the transports of enthusiasm and the paroxysms of terror with which that revolution was regarded by friends and foes. The dramatic grandeur of its circumstances; the expansive character it exhibited; the startling boldness of its doctrines and its aspirations; the eloquence, and heroism, and self-devotion, that mingled with and half redeemed its horrors, had all tended to awaken an almost delirious enthusiasm in Europe. Even in England, though the long-established free institutions and the strong aversion to everything French might have been deemed a sufficient barrier, the Government thought it necessary to put in motion all the long-disused engines of coercion to repress the new opinions. But in Ireland, where the ground-swell of agitation produced by the movement that had terminated in 1782 had not yet subsided, where the memory of the Volunteers was still fresh in every mind, where the traditions of past oppression and the spectacle of present abuses were alienating the people from England, while an affinity of character and an old debt of gratitude were drawing them to France, it is not surprising that the Revolution should have produced a deep and a lasting effect. As I have said, its adherents in Ireland were at first chiefly Protestants. What little republicanism existed in Ireland was mainly among the Presbyterians of Ulster. Wexford was the only county where the rebellion was distinctively Roman Catholic, and even there Bagenal Harvey, its leader, was a Protestant. Grattan and the Government both perceived the coming storm. The latter, in 1793, brought forward a Bill making those conventions which had hitherto proved the most powerful organs of public opinion illegal. Grattan, Curran, and Ponsonby warmly opposed the motion, but without success; and that Convention Act, which afterwards proved one of the greatest obstacles in O'Connell's course, became law. Grattan, on the other hand, urged the Government to grant those reforms by which alone rebellion could be averted, and the people to abstain from that violence which would imperil the existence of their constitution.

Ponsonby's Reform Bill was brought forward again, though without success, in 1794, and Grattan took the occasion to give a distinct outline of his policy. He desired 'that Ireland should improve her constitution, correct its abuses, and assimilate it as much as possible to that of Great Britain; that whenever Administrations should attempt to act unconstitutionally, but, above all, whenever they should tamper with the independence of Parliament, they should be checked by all means that the constitution justifies; but that these measures and this general plan should be pursued by Ireland with a fixed, steady, and unalterable resolution to stand or fall with Great Britain. Whenever Great Britain, therefore, should be clearly involved in war, Ireland should grant her a decided and unequivocal support, except that war should be carried on against her own liberty.'

At last it seemed as though better councils had prevailed. A large section of the Whigs, in consequence of the French Revolution, had deserted Fox, and had united themselves with Pitt, who, in order to ingratiate himself with his new allies, consented, after very considerable hesitation, to recall Lord Westmoreland, and to send over Lord Fitzwilliam as Lord-Lieutenant. Lord Fitzwilliam was one of the most important personages in the Whig party, an intimate friend of Grattan, and a warm and avowed supporter of Catholic emancipation. Such an appointment at such a moment could only be construed in Ireland in one way. Catholic emancipation was the pressing question of the hour. Pitt had early expressed himself in its favour. At a time when it was known to be in agitation he recalled a Viceroy who was opposed to it, and sent over one who was known to be its ardent friend. Lord Fitzwilliam was directed, indeed, not to bring it forward; but he had no instructions to oppose it, and was left, as he afterwards declared, a full discretion to deal with the question, if brought forward, as might seem to him advisable. Pitt himself asked an interview with Grattan, and stated to him the intended policy of the Government in a remarkable sentence, which was afterwards published by Grattan's son, on the authority of his father, and which there is no reason whatever for thinking inaccurately reported. Their intention was 'not to bring forward emancition as a Government, but if Government were pressed, to yield to it.'

Under these circumstances it appeared obvious that, if the dispositions of the Irish people and Parliament were favourable to emancipation, there was no obstacle to encounter. Lord Fitzwilliam landed in Ireland in December 1794, and was at once received with a most significant enthusiasm of loyalty. Petitions in unprecedented numbers poured in from the Catholics, asking for emancipation; and the great majority of the Protestants were unquestionably strongly in favour of it. Lord Fitzwilliam was afterwards able to represent to the King 'the universal approbation with which the emancipation of the Catholics was received on the part of his Protestant subjects;'[1] and in his letter to Lord Carlisle, after his recall, he described the state of

[1] See his letter to Grattan in Grattan's Life, by his son.

feeling in Ireland in terms which need no comment. It was a time, he wrote, 'when the jealousy and alarm which certainly at the first period pervaded the minds of the Protestant body exist no longer—when not one Protestant corporation, scarcely an individual, has come forward to deprecate and oppose the indulgence claimed by the higher order of Catholics—when even some of those who were most alarmed in 1793, and were then the most violent opposers, declare the indulgences now asked to be only the necessary consequences of those granted at that time, and positively essential to secure the well-being of the two countries.' Lord Fitzwilliam, in answering the addresses that were presented to him, used language which clearly intimated his sympathy with their cause; and such language, coming at such a time from the representative of the Sovereign, very naturally removed all doubts from the minds of the Catholics. In Parliament the almost universal feeling of the country was fully reflected. As on the occasion of Irish emancipation in 1782, extraordinary supplies were voted in testimony of the loyalty of the nation. Grattan, though without an official position, became virtually the leader of the Government; and the French party appeared to have almost disappeared. Grattan obtained leave to bring in an Emancipation Bill, with but three dissentient voices; and that Bill had been drawn up by him in concert with Lord Fitzwilliam and the Cabinet. It was understood that a Reform Bill would follow; and one of the most important leaders of the United Irishmen afterwards said that in that case their quarrel with England would have been at an end. The whole Catholic population were strung to the highest pitch of excitement. The Protestants were, for the most part, enthusiastically loyal; and the revolutionary spirit had almost subsided, when Pitt suddenly and peremptorily recalled Lord Fitzwilliam, and made the rebellion which followed inevitable.

The precise motives of this recall, which plunged Ireland into the agonies of civil war and threw back the Catholic question for thirty-four years, have been a matter of much controversy. Lord Fitzwilliam, in going to Ireland, thought it necessary to exercise his authority as chief governor by dismissing certain officials who were directly opposed to the policy he intended to pursue; and among these were two men of great influence: Cooke, the Secretary of War—who, a few years later, was put forward as the first advocate of the Union—and Beresford, a Commissioner of Revenue, who was at the head of one of the most powerful and most grasping families in Ireland. These measures were mitigated as much as possible; for Cooke was compensated by a pension of 1,200*l.* a year, and Beresford retained the whole of his official revenue; but Beresford, notwithstanding, went immediately to London; and he was supported in his complaints by the Chancellor, Lord Fitzgibbon, who was the favourite Minister of Pitt, and at the same time the bitterest enemy of the Catholics. To this influence the recall of Lord Fitzwilliam was ascribed, but though a reason, it was probably not the only one. It is scarcely probable that the dismissal of a subordinate Minister was the sole cause of a measure which plainly threatened the gravest and most disastrous consequences to the empire. The truth seems to be that Pitt was extremely jealous of his Whig colleagues, and afraid of their obtaining a predominant influence in the Cabinet. The King had declared his strong opposition to emancipation. The Minister would have found some difficulty with his Tory friends; and, although in the situation in which he then was it is almost certain that he could have carried the measure, he would have weakened and divided his party, and given the Whig element in his councils a considerable ascendency. He only sent over Lord Fitzwilliam with reluctance, and he probably hesitated and vacillated about the extent to which he was prepared to go. Personally he professed extremely liberal views about the Catholics, and he must have been quite aware of the danger of refusing their demands; but a careful examination will show that at every period of his career he sacrificed or subordinated political principles to party ends. But, besides these reasons, it is probable that he was already looking forward to the Union. The steady object of his later Irish policy was to corrupt and to degrade, in order that he ultimately might destroy, the Legislature of the country. Had Parliament been made a mirror of the national will—had the Catholics been brought within the pale of the constitution—his policy would have been defeated. Thus it was that a Minister who professed himself a warm friend of Catholic emancipation did more than any other English statesman to adjourn the solution of the question; that a Minister who began his career as the eloquent champion of Parliamentary reform resisted steadily every attempt to reform the most corrupt borough system in Europe; that a Minister whose political purity has been the theme of so many eulogists, was guilty in Ireland of a corruption before which the worst acts of Newcastle and Walpole dwindle into insignificance. The prominent part which Fitzgibbon and Cooke took in this transaction strengthens the probability that the contemplated Union had some influence over the decision of Pitt; and it is at least certain that the recall of Lord Fitzwilliam arrested a policy which would have made it at that time impossible. By raising the hopes of the Catholics almost to certainty, and then dashing them to the ground; by taking this step at the very moment when the inflammatory spirit engendered by the Revolution had begun to spread among the people; Pitt sowed in Ireland the seeds of discord and bloodshed, of religious animosities, and social disorganisation, which paralysed the energies of the country and rendered possible the success of his machinations. The rebellion of 1798, with all the accumulated miseries it entailed, was the direct and predicted consequence of his policy. Lord Fitzwilliam had solemnly warned the Government that to disappoint the hopes of the Catholics 'would be to raise a flame in the country that nothing but the force of arms could keep down.' Lord Charlemont, though on principle opposed to the Catholic claims, declared that the recall of Lord Fitzwilliam would be ruinous to Ireland, and foretold that by the following Christmas the people might be in the hands of the United Irishmen. The feelings of the nation were manifested with an intensity that had not been displayed since 1782. The shops of Dublin were closed; votes of confidence in the disgraced Lord-Lieutenant were passed unanimously by both Houses of Parliament, by most of the corporations in the kingdom, and by innumerable county meetings. His carriage was drawn to the water's edge by an enthusiastic crowd, while a violent riot marked the public entry of his successor. The belief in the possibility of obtaining reform by constitutional means speedily waned. A sullen, menacing disloyalty overspread the land, 'creeping,' in the words of Grattan, 'like the mist at the heels of the countryman.'

It was natural, and indeed inevitable, that it should be so. A large amount of discontent and agitation had previously existed, and it would have been very strange had it been otherwise. The past history of the country was not of a nature to make a contented people. The great armed movement of the Volunteers was still vivid in the memories of men, and the exclusion of three-fourths of the nation from the highest privileges of the constitution, the profoundly corrupt condition of Parliament, and the systematic misapplication of official patronage, were most legitimate causes of discontent. Still the disloyalty was probably less than at the present moment, and it might most easily have been allayed. Had the Government thought fit to adopt the policy of Grattan—had they determined 'to combat the wild spirit of democratic liberty by the regulated spirit of organised liberty, such as may be found in a limited monarchy with a free Parliament,' there can be little doubt that they would have succeeded. The landlords, the Parliament, the overwhelming majority of the Episcopalian Protestants, the Constitutional Liberals who followed Grattan and Charlemont, were intensely loyal. The priests in Ireland, as elsewhere, looked with horror upon the Revolution, and upon the doctrines that inspired it. The mass of the Catholics were, no doubt, considerably and most naturally discontented, but their leaning was strongly towards authority, and the contagion of the disloyal spirit that was agitating the Presbyterians of the north did not seriously affect them till the recall of Lord Fitzwilliam. On this point we have the evidence of the most competent of witnesses: the three leaders of the United Irishmen, whose memoir I have before cited. 'Whatever progress this united system had made among the Presbyterians of the north,' they say, 'it had, as we apprehend, made but little way amongst Catholics throughout the kingdom, until after the recall of Lord Fitzwilliam.' The conduct of the people in 1782, and their conduct on the arrival of Lord Fitzwilliam, attested sufficiently how easily all classes might have been rallied round the throne, and though some agitators would always have remained, they would have been reduced to impotence, if not to silence, by Catholic emancipation and a moderate measure of Parliamentary reform. Considering the past history of the country, and the inflammatory elements that were abroad in Europe, Ireland in 1795 was singularly easy to govern, had it been governed honestly and by honest men. But it was not in human nature that the loyalty of the Catholics should survive the administration of Lord Fitzwilliam. Their hopes had been raised to the highest point; the language and demeanour of the representative of the Sovereign had been equivalent to a pledge that they would be relieved of their disqualifications; they could point with pride to their perfect loyalty for the space of a hundred years, in spite of the penal laws, of the rebellions of 1715 and of 1745, and of the revolt of the colonies; they had won to their cause the immense majority of their Protestant fellow-countrymen, and had advanced to the very threshold of the constitution, when the English Minister interposed to blight their prospects, and exerted all the influence of the Government against them.

It has been suggested, by a distinguished modern apologist for Pitt, that it would perhaps have been impossible to carry the measure through the Irish Parliament; or that, at least, it could only have been carried after a prolonged and violent conflict, that would have shaken the nation to the centre. The fact that the House almost unanimously gave permission for the Bill to be brought in does not, it is truly said, necessarily imply that it would have passed it in its more advanced stages; and when, soon after, the Government opposed the Bill, it was rejected by a large majority. The answer to this theory is very short. No Irish writer or speaker of the time questioned, as far as I am aware, the power of the Government to carry the Bill. The weight which the Administration possessed through the borough influence in 'the Irish Parliament was almost absolute. In a few cases a strong popular feeling was able to defeat it, but in no case had the Government any serious difficulty when the popular sentiment was on their side. That the general feeling of the people was in favour of emancipation is perfectly unquestionable. That the Parliament would readily have yielded to that feeling is decisively proved by its conduct in 1793. The Bill carried in that year, which conferred the elective franchise on the Catholics, was, as I have said, far more important than a Bill for allowing them to sit in Parliament, for it transferred a far larger amount of real political power, and rendered the Parliamentary disqualification utterly untenable. The Bill of 1793 was carried without difficulty through Parliament, and there is not a shadow of a reason for believing that it would have been more difficult to carry the Emancipation Bill of 1795. In the emphatic words of Lord Fitzwilliam, the disqualifications that were retained in the Act of 1793 'gave satisfaction to none, and caused discontent to many. The Protestants regarded these exceptions with total indifference. The Catholics looked on them as signs of suspicion and degradation.'[1] There may, perhaps, be some difficulty in deciding on whose head the blame of the failure of Lord Fitzwilliam's viceroyalty should rest; but it is at least very clear that the real obstacle to Catholic

[1] Protest in the House of Lords.

emancipation was not in Ireland, but in England. Few facts in Irish history are more certain than that the Irish Parliament would have carried emancipation if Lord Fitzwilliam had remained in power, and that the recall of that nobleman was one of the chief causes of the rebellion of 1798. Lord Fitzwilliam, on his return, demanded in the House of Lords explanations of the motive of his recall, and was supported by the Duke of Norfolk, but his demand was refused. He entered a protest against this refusal, in which he stated that he found Catholic emancipation to be 'ardently desired by the Roman Catholics, to be asked for by very many Protestants, and to be cheerfully acquiesced in by nearly all.'

After this event the days of the Irish Parliament were but few and evil. Three or four times Grattan brought forward the Catholic and the Reform questions, but the Government continually refused to yield, and the revolutionary tide surged higher and higher. At last, on the eve of the rebellion, he gave up his seat in Parliament, and retired into private life. He had found it wholly impossible to cope with the Government during that period of panic. He could not sympathise with the party who were appealing to arms, nor yet with those who had driven them to disloyalty. He was guided, too, in a great measure, by the example of Fox, who, when he found his party hopelessly reduced, had retired from the debates; but, unlike Fox, he resigned his seat when he abstained from parliamentary business.

If it were not for the wretched condition of the country, it would have cost him comparatively little to retire from active life; for he possessed all the resources of happiness that are furnished by a highly cultivated intellect, by the most amiable of dispositions, and the attachment of innumerable friends. All accounts concur in representing him in private life as the simplest and most winning of mortals. The transparent purity of his life and character, a most fascinating mixture of vehemence and benevolence, a certain guilelessness of appearance, and a certain unconscious oddity, both of diction and gesture, gave a peculiar charm and pungency to his conversation. Like his speeches, it was tesselated with epigram and antithesis, full of strokes of a delicate, original, and laconic humour, of curiously minute and vivid delineations of character, of striking anecdotes, admirably though quaintly told. He had seen and observed much, and he possessed a rare insight into character and a great originality both of thought and of expression. He delighted in music and poetry, and his love of nature amounted to a passion, and continued unabated during every portion of his life. In one of the letters of Horner there is a charming description of the enthusiasm with which, when an old man, he left London to visit a county which was famous for its nightingales, in order that he might enjoy the luxury of their song. There was about him so much greatness and so much goodness, that he rarely failed to win the love and the veneration of those who came in contact with him, but also so much oddity that he usually provoked a smile. With much mild dignity of manner and great energy of intellect, he combined an almost childish simplicity and freshness of character. No schoolboy enjoyed with a keener zest a day's holiday in the country; and Curran, who delighted in mimicking his singularities, described him conducting a controversy about the respective merits of two pumps, with an intensity of earnestness and a measured gravity worthy of a great political contest. It is a fine saying of

Coleridge that in men of genius the matured judgment of the man is combined with the delicacy of feeling and the susceptibility of impressions of the child, and it needs but little acquaintance with literary biography to perceive that these last elements almost invariably enter into the composition of really great men. It is scarcely less true of the temple of genius than of the temple of Christianity, that he who would enter in must become as a little child.

It does not fall within the province of the present work to paint the rebellion of 1798. Public opinion had but little scope during a period of military law and of mob violence, and the historians of the two countries may well let the curtain fall over a scene that was equally disgraceful to both. The man who at that time occupied the first position in the public mind was, beyond all question, Curran. Seldom has Ireland produced a patriot of more brilliant and varied talents; and although there were grave defects in his private character, his public life was singularly unblemished, and there are few of his contemporaries who inspire a feeling so much akin to affection. Rising from a position of the deepest humility, he early attracted public attention as a poet of no mean promise—a wit of almost the highest order—and an orator who might compare with the greatest of his countrymen. If his speeches, like those of most lawyers, are somewhat lax and inaccurate in their style; if they do not exhibit great depth of thought or great force of reasoning, they are characterised at least by a musical flow that delights even in an imperfect and uncorrected report, and by a power of pathos, of imagination, and of humour that was equalled by none of his contemporaries. A member of a profession where all promotion depended on the

Government, and was then given from political motives, he was never guilty of abandoning a principle or swerving from a public duty. He exhibited the most chivalrous courage in one of the worst periods of judicial intimidation, and the most perfect disinterestedness in one of the worst periods of judicial corruption. At the very beginning of his career he signalised himself by volunteering to defend an old priest who had been maltreated by a Protestant nobleman, and whose cause no other member of the Bar was willing to adopt. Lord Clare drove him from the Court of Chancery by continual evidences of dislike. Lord Carleton hinted to him that he might lose his silk gown for his defence of the United Irishman Neilson. During one of his speeches he was interrupted by the clash of the arms of an angry soldiery, and more than once he had to dread those political duels by which dullness so often revenged itself upon genius. In his famous speech for Hamilton Rowan he could adopt almost without alteration the exordium of Cicero's defence of Milo, but, unlike Cicero, the attempts at intimidation that he described only served to stimulate his eloquence. And yet this man, before whose sarcasm and invective corrupt judges and perjured witnesses so often trembled; this man, on whose burning eloquence crowded and sometimes hostile courts hung breathless with admiration till the shadows of evening had long closed in, was in private life the most affable, the most gentle, the most unassuming of friends. The briefless barrister, the young man making his first essays of ambition, the bashful, the needy, and the disappointed, ever found in him the easiest of companions, and acknowledged with delight that his social qualities were as fascinating as his eloquence.

Like his great contemporary Erskine, he never obtained in Parliament a position corresponding to that which he held at the Bar; but his Parliamentary career, if not very brilliant, was at least eminently consistent and disinterested. He made his maiden speech in favour of Flood's Reform Bill, and he took part in almost every subsequent effort to purify the Parliament, to emancipate the Catholics, to reduce the pensions, to ameliorate the criminal code, and to prevent the introduction of military law. He laboured with especial earnestness, though without success, to assimilate the law of treason in Ireland to that of England, by which two witnesses were necessary for a capital conviction; and if he had succeeded he would have prevented some of the most scandalous scenes that disgraced the subsequent prosecutions. In all the great trials of '98 he was the counsel for the prisoners, and his eloquence proved fully equal to the occasion. His finest effort is his defence of Hamilton Rowan, which has been styled by the first of our oratorical critics [1] the most eloquent speech ever delivered at the Bar, but which is said to owe a great deal of its pre-eminence to the fact that it was better reported than his other speeches. It was on that occasion that he broke into his eloquent and well-known justification of the principle of 'universal emancipation,' which had been asserted by the United Irishmen, and denounced by the Crown officers as treasonable. 'I speak in the spirit of the British law, which makes liberty commensurate with and inseparable from the British soil; which proclaims even to the stranger and the sojourner, the moment he sets his foot on British earth, that the ground on which he treads is holy and consecrated by the genius of universal emancipation. No matter in what language his doom may have been pronounced—

[1] Lord Brougham, in his defence of Hunt.

no matter what complexion, incompatible with freedom, an African or an Indian sun may have burnt upon him —no matter in what disastrous battle his liberty may have been cloven down—no matter with what solemnities he may have been devoted upon the altar of slavery —the first moment he touches the sacred soil of Britain the altar and the god sink together in the dust; his soul walks abroad in its own majesty; his body swells beyond the measure of his chains that burst from around him, and he stands redeemed, regenerated, and disenthralled by the irresistible genius of universal emancipation.'

The rebellion of '98 was at last suppressed, and the Ministers determined to avail themselves of the opportunity to annihilate the Irish Parliament. The notion of a Union had been more than once propounded in both countries. Cromwell had summoned Irish members to the Parliament in Westminster. Many eminent writers had advocated a Union—among others, Sir W. Petty, Dean Tucker, and Adam Smith; and about the time of the Union with Scotland strong efforts were made by Irish politicians to effect it. In 1703 there was a certain movement in this sense; and in 1709 the Irish House of Lords—though apparently without the concurrence of the House of Commons—petitioned Lord Wharton, the Lord-Lieutenant, to use his good offices to procure for Ireland a Union like that between England and Scotland.[1] The reply of the Lord-Lieutenant, however, was exceedingly discouraging; and from this time the question seems to have slept till 1759, when a report was current that such a measure was contemplated; and so unpopular was the project, that the Dublin mob seized a number of the members, and made them swear that they would vote against it. In 1786 we find

[1] See Lord Mountmorres' 'Historical Dissertation on the Irish Parliament,' p. 47.

Charlemont writing to Flood: 'The English papers have lately been infested with the idea of a Union, but except from them I know nothing of it; neither can I suppose it possible that such a notion can have entered into the heads of our present Administration. When we had no constitution the idea was scarcely admissible; what, then, must it be now?' Wilberforce, on one occasion, observed that it would be a good measure, but impracticable, for the people would never consent. Dr. Johnson said to an Irish gentleman, 'Do not unite with us; we would unite with you only to rob you.' The Lord-Lieutenant was Lord Cornwallis, in whose published correspondence we can trace very clearly the progress of the design; but the principal agent of the Government in corrupting the Legislature was the Chief Secretary, Lord Castlereagh. In the November of 1798 we find the following curious notice of this appointment in one of Lord Cornwallis's letters to the Duke of Portland: 'Lord Castlereagh's appointment gave me great satisfaction; and although I admit the propriety of the general rule, yet, as he is so very unlike an Irishman, I think he has a great claim to an exception in his favour.' In the same month we find Lord Castlereagh writing to Mr. Wickham: 'The principal provincial newspapers have been secured, and every attention will be paid to the press generally.' The public were prepared by a pamphlet in favour of a Union written by the Secretary Cooke, which elicited a multitude of answers, the ablest being those of Bushe and Jebb. Parnell and Fitzgerald, who refused to acquiesce in the designs of the Government, were dismissed from office; and in 1799, after what was considered a sufficient distribution of bribes and promises,[1] the measure was introduced.

---

[1] The following notice in the Cornwallis Letters concerning Archbishop Agar is amusingly characteristic. It is in a letter from Lord

The period was in many respects very favourable to the attempt. In the House of Lords there was no serious opposition to be apprehended. Peerages in Ireland had long been granted almost exclusively with a view to ensure ministerial influence, and Pitt had surpassed all his predecessors in the lavish audacity of his creations. The bishops, who were absurdly numerous in proportion to their flocks, were, with two exceptions, docile and obsequious; and by ennobling most borough-owners who consented to send servile members to the House of Commons, the Minister was able, with an economy of corruption, to degrade two Houses. On all ordinary questions he could secure a majority in the

Cornwallis to the Duke of Portland, in July 1797: 'It was privately intimated to me that the sentiments of the Archbishop of Cashel were less unfriendly to the Union than they had been, on which I took an opportunity of conversing with his Grace on the subject, and, after discussing some preliminary topics respecting the representation of the Spiritual Lords, and the probable *vacancy of the see of Dublin*, he declared his great unwillingness at all times to oppose the measures of the Government, and especially on a point in which his Majesty's feelings were so much interested, to whom he professed the highest sense of gratitude, and concluded by a cordial declaration of friendship.' Dr. Agar was made a viscount in 1800, Archbishop of Dublin in 1801, and Earl of Normanton a few years later. He tried very hard to obtain the Primacy of Ireland, but the Government refused to relax their rule that no Irishman should hold that place. However, Lord Cornwallis writes: 'His Grace had my promise when we came to an agreement respecting the Union that he should have a seat in the House of Lords for life' ('Cornwallis Correspondence,' iii. pp. 160–209). Archbishop Agar was also remarkable for the zeal with which he advocated sanguinary measures of repression during the rebellion of 1798 (Grattan's Life, vol. iv. p. 390); for the large fortune which he made by letting the Church lands on terms beneficial to his own family ('Castlereagh Correspondence,' vol. ii. p. 71); and for having allowed the fine old church at Cashel to fall into ruins, and built in its place a cathedral in the worst modern taste, which he ordered to be represented on his tomb (Stanley's 'Westminster Abbey,' p. 324). There is an extremely eulogistic inscription to his memory in Westminster Abbey, and a fine bas-relief representing the angels bearing the mitre to the saintly prelate.

Lower House, and he had been for many years increasing the number of placemen. In quiet times a storm of popular indignation would have made a Union impossible, but at the time when the measure was brought forward the country was prostrate and paralysed after the great rebellion. Resistance was impossible, and there was much to predispose men to a Union. The civil war which the policy of Pitt had produced degenerated in Wexford, and in part of the south, into a merciless struggle of races and creeds, disgraced on both sides by the most atrocious cruelties. The Protestants passed into that condition of terrified ferocity to which ruling races are always liable when they find themselves a small minority in the midst of a fierce rebellion. 'The minds of the people,' wrote Lord Cornwallis, after the suppression of the revolt, 'are now in such a state that nothing but blood will satisfy them.' 'Even at my table, where you will suppose I do all I can to prevent it, the conversation always turns on hanging, shooting, burning, and so forth; and if a priest has been put to death the greatest joy is expressed by the whole company.' The Catholics were equally sanguinary. A prominent rebel, who was executed on Vinegar Hill, and whose confession is preserved in the 'Castlereagh Correspondence,' gives a graphic account of their proceedings: 'Every man that was a Protestant was called an Orangeman; and every one was to be killed, from the poorest man in the country. Before the rebellion I never heard there was any hatred between Catholics and Protestants; they always lived peaceably together. I always found the Protestants better masters and more indulgent landlords than my own religion. During the rebellion I never saw any one interfere to prevent murder but one Byrne, who saved a man.'

Under these circumstances it would not have been surprising if the Protestants, terrified at the fierce elements that were surging around them, should have welcomed any political combination that, by identifying them more completely with a powerful Protestant nation, might increase their strength; or if the Catholics should have accepted with equal delight a measure that withdrew them from the immediate tyranny of their enraged fellow-countrymen. But beside these considerations, political inducements of a more special kind were persistently and adroitly employed. One of the strongest wishes of the Irish Catholics was naturally to be freed from their political disqualifications. One of the most serious objections in the eyes of the Irish Protestants to Catholic emancipation was that it might prove fatal to the permanence and security of the Established Church in Ireland. The Ministers and the ministerial writers argued that a Union would lead to the immediate consummation of the wishes of the Catholics, and that it would at the same time place the Establishment beyond all possibility of danger. Catholic emancipation, as we have already seen, had been looked upon very favourably by the Irish Protestants; but Pitt having suffered Lord Fitzwilliam to amuse the Irish people by the prospect, had blighted their hopes by recalling him, and thus produced the rebellion. Irish opinion had greatly deteriorated under the influence of the events that followed, and sectarian animosity was much stronger in 1799 than in 1795; but still the passing of the measure depended upon the attitude of the Government. With their assistance it was easy. In the face of their opposition, and with an unreformed Parliament, it was impossible. Being thus the practical arbiters of the question, they determined to employ it as a means of compelling the

Catholics to support the Union. Pitt himself—whose political speculations were almost always large and liberal—wished to give Catholic emancipation with the Union, and would certainly have done so if he could have accomplished the object without in any degree, diminishing or endangering his political ascendency. His great aim, however, was not to emancipate the Catholics, but to make them believe that he was going to do so, and thus to bribe them to support the Union. The enterprise was a difficult one, for Lord Clare and some of the other chief advocates of the Union were very hostile to the Catholics; and the Minister desired to enlist in his support all the anti-Catholic elements in the country. The plan, therefore, of coupling the Union with favours to the Catholics was abandoned; although Pitt wrote to Lord Cornwallis in November 1798, that Mr. Elliot—who was one of his chief authorities on Irish matters—thought that a Union, accompanied by Catholic emancipation, ought to be, and might easily be, accomplished; and although Pitt himself noticed at the same time that all the Irish he had seen were in favour of 'a provision for the Catholic clergy, and of some arrangement respecting tithes.'[1] Another course of proceeding was resolved upon. The leading Catholics were to be privately assured that though Government would oppose emancipation as long as the Irish Parliament existed, they desired to carry it if the Union was effected. In the autumn of 1799 Cornwallis directed Castlereagh to inform the English Government that the Union could not be carried if the Catholics were in active opposition, and that their attitude on the question depended mainly upon their hopes of emancipation. He added that friends of the Government had already

[1] Stanhope's 'Life of Pitt,' vol. iii. p. 161.

produced a favourable impression by exciting those hopes, and he desired to know how far he might pursue that course. A Cabinet having been hastily summoned, Castlereagh informed him, as the result, that the Ministers who composed it were unanimously in favour of the principle of emancipation; that they apprehended considerable repugnance to the measure in many quarters, and particularly in the highest; that they declined to give an express promise, partly because it would embarrass them in their negotiations with the Protestants, and partly because it was not right that such claims should be made a matter of mere bargain; but that, as far as the sentiments of the Cabinet were concerned, the Lord-Lieutenant was fully authorised to solicit the Catholic support.[1] This pretended unanimity, in fact, did not exist, for when the question was formally brought forward in the Cabinet in 1801, it appeared that no less than five of its members were opposed to emancipation;[2] but of this the Catholic leaders could know nothing. They were probably aware that the King was hostile to emancipation, but they could not know that both in 1795 and 1798 he had distinctly declared that his objections to it were insuperable,[3] and that the overtures made to them were made with a perfect knowledge of his sentiments, without any attempt to learn how far they might be modified,[4] or any determination to exert the full ministerial

---

[1] 'Cornwallis Papers,' vol. iii. p. 326.

[2] Stanhope's 'Life of Pitt,' vol. iii. p. 273.

[3] See the very remarkable paper drawn up by the King in 1795, in Campbell's 'Lives of the Chancellors,' vol. viii. pp. 173-175; and his letter to Pitt in 1798, in Stanhope's 'Life of Pitt.' iii. Append. xvi.

[4] It is charitable to suppose that Pitt really hoped to carry emancipation by forcing the hand of the King after the Union was carried. Mr. Adolphus, who had much private information of the proceedings at court, says: 'The assurance was given to the Irish Catholics without

power in their favour. They only knew that the chief Irish representatives of one of the strongest governments that ever existed in England represented the Cabinet as unanimously in favour of emancipation, and on that ground solicited their support. Government influence alone had defeated emancipation in 1795. They were told that the Government objection to it would be obviated by a Union, and they inferred that by carrying the Union they were carrying their cause. The great object was to hold out hopes sufficient to secure Catholic support or neutrality without committing the Government to a distinct pledge; and this end was most dexterously accomplished. A few sentences written by Lord Castlereagh in 1799 explain the calculation that was made. 'The Catholics,' he says, ' if offered equality without a Union, will probably prefer it to equality with a Union; for in the latter case they must ever be content with inferiority, in the former they would probably by degrees gain ascendency. . . . Were the Catholic question to be now carried, the great argument for a Union would be lost, at least as far as the Catholics are concerned; it seems, therefore, more important than ever for Government to resist its adoption, on the ground that without a Union it must be destructive; with it, that it may be safe.'[1]

While this powerful inducement was offered to the Catholics, another and almost equally strong one was offered to the Protestants. Both Flood and Charlemont, as I have already stated, had objected to Catholic emancipation on the ground that it would lead to the

the King's privity, and with a full knowledge of his sentiments upon the subject, in the hope that his Majesty, after the Union had taken place, seeing that Catholic emancipation was indispensable, would agree, however reluctantly, to that measure.'—*History of England*, vol. vi.

[1] 'Castlereagh Correspondence,' vol. ii. p. 140. There are other passages to the same effect in the Correspondence.

disestablishment and disendowment of the Established Church, and the advocates of Catholic emancipation had always rejected the prophecy with indignation. By the Union it was maintained that the Church would be placed in absolute security, and this security was one of the special grounds upon which the Protestants were urged to support it. It was of two kinds. The Act of Union was looked upon as a treaty by which the Irish Parliament consented on certain conditions to surrender its separate existence, and one article of that Act, inserted by the desire of Archbishop Agar, stipulated for the preservation of the Establishment as ' an essential and fundamental part of the Union.' Besides this, the Church being placed under the protection of a Legislature which was likely at all times to remain mainly Protestant, it was imagined that no serious danger could menace it. The stress laid upon these considerations by the Government advocates of the Union was very great. 'With the Union,' wrote the Secretary Cooke, 'Ireland would be in a natural situation; for, all the Protestants of the empire being united, she would have the proportion of fourteen to three in favour of her Establishment, whereas at present there is a proportion of three to one against it.' ' So long as the separation shall continue,' said Castlereagh, ' the Church of Ireland will ever be liable to be impeached upon local grounds. Nor will it be able to maintain itself effectually against the argument of physical force. But when once completely incorporated with the Church of England, it will be placed upon such a strong and natural foundation as to be above all apprehensions and alarms.'

It is a curious enquiry how far public opinion was influenced by these considerations. The last which I have mentioned appears to have had extremely little effect. Clare, Duigenan, and the bishops, it is true, were ardent advocates of the Union, but it appears tolerably certain that no considerable section of Protestants of any class outside Parliament concurred in their view. The Orangemen were decidedly hostile, and the utmost that could be obtained of them was that they would not act in their corporate capacity in opposition to it. The Established Church has played an important part in the history of the Union, but it was at a much later period. The conviction that repeal would be followed by disestablishment was one of the reasons that arrayed the great majority of the Protestants in hostility to O'Connell, and the connection between the two measures was clearly recognised. When Lord John Russell in 1835 was endeavouring to apply a very small part of the Irish Church revenues to secular purposes, Mr. Gladstone, in a speech of consummate eloquence, denounced the policy of the Whig leader, and predicted the consequences that might flow from it. 'The noble Lord invited them to invade the property of the Church in Ireland. He (Mr. Gladstone) considered that they had abundant reasons for maintaining that Church, and if it should be removed he believed that they would not be long able to resist the repeal of the Union.'[1]

With reference to the Catholics, however, the case is somewhat different. Those of Dublin, indeed, took an active and emphatic part against the Union, and the great majority of them throughout the country were probably either hostile or indifferent to it.[2] There was, however, unquestionably a real and considerable

---

[1] Hansard, 3rd ser. vol. xxvii. p. 513.
[2] See the complaint of Lord Cornwallis (Jan. 31, 1800), that the Catholics were 'joining the standard of opposition.'—*Cornwallis Correspondence.*

Catholic party in its favour, guided with remarkable skill and energy by Troy, the Archbishop of Dublin, comprising, among other prelates, the Archbishops of Tuam and of Cashel,[1] and favoured by an important section of the Catholic aristocracy. Corry, the Chancellor of the Exchequer, the most violent opponent of Grattan in the Union debates, won his seat at Newry through the unanimous support of the Catholic voters.[2] Considered in the light of subsequent history, it is a curious fact that the Union was least unpopular in the Province of Munster and in the towns of Cork and Sligo, and that some of the Catholic priests were among the most active agents in procuring signatures to addresses in its favour.[3]

It is scarcely an exaggeration to say that the whole unbribed intellect of Ireland was opposed to it. Almost the only man of considerable talent in the Ministerial ranks was Fitzgibbon, who held the office of Lord Chancellor, and had obtained the Earldom of Clare. He was a ready and powerful debater, and a man of great personal courage and force of character, but he never appears to have been suspected even by his friends of any patriotic feelings, his intellect was narrow and intolerant, and his temper ungovernably violent. He had been at one time considered a Liberal, and owed his promotion in a great degree to Grattan, whom he afterwards attacked with the utmost virulence. Like many Irishmen of a later time, he had the habit of constantly depreciating and vilifying his country—' our damnable country,' as he described it in a letter to Lord Castlereagh—and he was a bitter enemy of the Catholics. He was remarkable for an arrogance of

---

[1] See the 'Castlereagh Correspondence,' vol. ii. pp. 344-348.
[2] Ibid. p. 168.
[3] Ibid. pp. 26, 328, 348, 349, 400.

tone, which in debate is said sometimes to have almost verged upon insanity, and for the reckless manner in which he displayed his personal antipathies upon the Bench; and he scandalised the Irish Parliament by the perfect frankness with which he justified a policy of corruption, and the English House of Lords by his apology for the use of torture against the rebels of 1798. Probably no Irishman of his generation was so hated, and when he died the popular delight broke out (as it afterwards did in England at the death of Castlereagh) in a kind of hideous carnival around his coffin. He was, however, quite capable of generous actions, and showed on one or two occasions real humanity towards State prisoners in 1798; and his rare skill in stating a case, and his indomitable courage in meeting opposition, made him extremely useful to the Ministry. For many years he was almost absolute in the House of Lords, and after Lord Castlereagh he contributed most to passing the Union. It is, however, curiously illustrative of the tortuous skill with which the Administration of Pitt was conducted, that Clare, when apparently the very leader of the Ministerial party, was kept in complete ignorance of the secret overtures that were made to the Catholic prelates, and of the intentions of the Minister to make the Union the prelude to emancipation.[1]

The Irish Bar was at this time peculiarly rich in

[1] Lord Holland says: 'Lord Hobart afterwards assured me that both he and Lord Clare had been deceived by Mr. Pitt, and that he would have voted against the Union had he suspected at the time that it was connected with any project of extending the concessions already made to the Irish Catholics. The present Lord Clare's report of his father's views of the whole matter tallies with this account of the transaction.'—*Memoirs of the Whig Party*, vol. i. p. 162. See, too, on the indignation of Lord Clare at what he called the 'deception' that was practised on him, the 'Castlereagh Correspondence,' vol. iv. pp. 47, 50.

talent, and one of the first objects of the Government was to corrupt it. To a certain extent the lawyers had undoubtedly a professional interest in keeping the Parliament in Dublin, but, on the other hand, all promotions were in the hands of the Government, and every power which the Government possessed was unscrupulously strained. It was certain, beyond all reasonable doubt, that the overwhelming majority of the people of Ireland were opposed to the destruction of their national Parliament, but it was necessary to create some semblance of popular opinion on the other side, and accordingly Castlereagh began his campaign by drawing 5,000*l.* from the secret service fund, and expended the greater part of it in bribing young lawyers to write pamphlets in favour of a Union. The vehement part which the Chancellor took in advocating the Union had naturally an influence upon the Bar. All officials who held any office under Government were rigorously expelled if they would not support it, while, on the other hand, crowds of unprincipled and incompetent men were promoted to high legal honours for defending it. The immobility of the judges having been conceded shortly after the emancipation of Parliament, and the penal laws having been for the most part abrogated, there was every reason to believe that by a just and upright policy the antipathy to law which had become so deeply ingrained in the Irish character might have been gradually removed. The judicial promotions that followed the Union directly and powerfully strengthened it. Lawless men are not likely to learn to reverence the law when it is administered by officials whose positions are notoriously the reward of their political profligacy.

The conduct of the Irish lawyers at this time was on the whole eminently noble. In spite of the lavish corruption of the Ministers, the great body remained firm to the anti-Ministerial side, and both in public meetings and in Parliament they were the most ardent opponents of the Union. Nor does there appear in this respect to have been any considerable difference between Whigs and Tories, or between Protestants and Catholics. When the measure was first propounded a great meeting was held under the presidency of Saurin, one of the ablest of the Tory lawyers, and was attended by all the leading lawyers of all sides, and at this meeting a resolution condemning the proposed Union was carried by 166 to 32. At the end of 1803 there were only five members of the minority who had not received appointments from Government. In Parliament the speeches of Plunket and of some of his legal colleagues were masterpieces of powerful reasoning, and should be studied by all who desire to know the light in which the measure appeared to some of the most disciplined intellects in the community. It would, indeed, be scarcely possible to find in the whole compass of Parliamentary eloquence speeches breathing a more intense bitterness. 'I will make bold to say,' said Plunket, 'that licentious and impious France, in all the unrestrained excess which anarchy and atheism have given birth to, has not committed a more insidious act against her enemy than is now attempted by the professed champion of the cause of civilised Europe against her friend and ally in the time of her calamity and distress—at the moment when our country is filled with British troops—when the loyal men of Ireland are fatigued and exhausted by their efforts to subdue the Rebellion—efforts in which they had succeeded before those troops arrived—whilst the Habeas Corpus Act is suspended—whilst trials by court-martial are carrying on in many parts of the kingdom—whilst the people are taught to think they have no right to meet or deliberate, and whilst the great body of them are so palsied by their fears or worn down by their exertions that even this vital question is scarcely able to rouse them from their lethargy—at a moment when we are distracted by domestic dissensions—dissensions artfully kept alive as the pretext of our present subjugation and the instrument of our future thraldom.' 'For centuries,' said Bushe, 'the British Parliament and nation kept you down, shackled your commerce and paralysed your exertions, despised your characters and ridiculed your pretensions to any privileges, commercial or constitutional. She has never conceded a point to you which she could avoid, nor granted a favour which was not reluctantly distilled. They have been all wrung from her like drops of blood, and you are not in possession of a single blessing (except those which you derived from God) that has not been either purchased or extorted by the virtue of your own Parliament from the illiberality of England.' The language of Saurin was still stronger. 'If a legislative Union,' he said, 'should be so forced upon this country against the will of its inhabitants it would be a nullity, and resistance to it would be a struggle against usurpation, and not a resistance against law. You may make it binding as a law, but you cannot make it obligatory on conscience. It will be obeyed as long as England is strong, but resistance to it will be in the abstract a duty, and the exhibition of that resistance will be a mere question of prudence.' 'When I take into account,' said Burrowes, 'the hostile feelings generated by this foul attempt, by bribery, by treason, and by force, to plunder a nation of its liberties in the hour of its distress, I do not hesitate to pronounce that every sentiment of affection for Great Britain will perish if this measure pass, and that, instead of uniting the nations, it will be the commencement of an era of inextinguishable animosity.'

The combined exertions of almost all the men of talent and of almost all the men of pure patriotism in the Parliament were successful in 1799. The Government Bill was defeated by 109 to 104, and the illumination of Dublin attested the feeling of the people. The national party did all that was in their power to secure their triumph, for they foresaw clearly that the struggle would be renewed. Ponsonby brought forward a resolution pledging the House to resist every future measure involving the principle it had condemned, but he was compelled eventually to withdraw it. Mr. Dobbs, a lawyer of some talents and the purest patriotism, but whose influence was impaired by an extraordinary monomania on the subject of prophecy,[1] brought forward a series of measures for the purpose of tranquillising the country, comprising Reform, Catholic Emancipation, and the payment of the priests; but the Government was again successful, and the shadow of the coming year fell darkly on every patriotic mind.

These gloomy forebodings were soon verified. After a series of measures of corruption which I shall presently

---

[1] He believed that Armagh is Armageddon. The Irish, it appears, of Armagh is Armaceaddon; *c* and *g* are interchangeable letters, and thus, by contraction, we should have Armageddon. Armaceaddon means the hill of the prophet; and some 'eminent Hebrew scholar' considered that Armageddon meant much the same. Mr. Dobbs also considered that the 'white linen' in the Apocalypse alluded to the linen trade in Ireland, the sea of glass to its insular position, the harps borne by the angels to its national arms, and that the Giant's Causeway was the Stone of Daniel. He wrote two books, 'A Short View of Prophecy' and 'A Universal History,' both in letters to his son. Unlike most persons who indulge in these eccentric opinions, he was as liberal as he was patriotic, and was selected by Grattan to carry the resolutions in favour of the repeal of the penal laws to the Volunteers at Dungannon.

describe, the Union was again introduced, and this time with success. Grattan was suffering from a severe illness. His strength was completely prostrated, and he was not in a fit condition for the most moderate exertion, far less for a great political contest. In his country's extremity, however, it was not fitting that he should be absent from her councils, and he accordingly procured his election for Wicklow, and entered the House during the debate. He wore the uniform of the Volunteers. He was so feeble that he could only walk with the assistance of two friends, and his head hung drooping upon his chest, but an unwonted fire sparkled in his eye, and the flush of deep emotion mantled his cheek. There was a moment's pause, an electric thrill passed through the House, and then a long wild cheer burst from the galleries. Shortly afterwards he rose to speak, but his strength failed him, and he obtained leave to address the House sitting. Then was witnessed that spectacle, among the grandest in the whole range of mental phenomena, of mind asserting its supremacy over matter—of the power of enthusiasm and the power of genius nerving a feeble and an emaciated frame. As the fire of oratory kindled —as the angel of enthusiasm touched those pallid lips with the living coal—as the old scenes crowded on the speaker's mind, and the old plaudits broke upon his ear, it seemed as though the force of disease was neutralised, and the buoyancy of youth restored. His voice gained a deeper power, his action a more commanding energy, his eloquence an ever-increasing brilliancy. For more than two hours he poured forth a stream of epigram, of argument, and of appeal. He traversed almost the whole of that complex question—he grappled with the various arguments of expediency the Ministers had urged; but he placed the issue on the highest of grounds. 'The thing he proposes to buy is what cannot be sold—liberty.' When he at last concluded, it must have been felt by all his friends that if the Irish Parliament could have been saved by eloquence it would have been saved by him. He had been for some time vehemently denounced in Parliament, and Corry now attempted to crush him by a violent attack. Grattan, however, treated his adversary with contemptuous silence till the assault had been three times repeated, when he terminated the contest by a very brief but most crushing invective, and a duel, in which Corry was wounded, was the result.

It was soon evident, however, that no eloquence and no arguments could save the constitution of Ireland. In division after division Grattan was defeated, and he saw with an ineffable anguish the edifice which he had done so much to construct sinking into inevitable dissolution. Night after night the contest was vainly prolonged with a feverish and impassioned earnestness. Yet, even at that period, hope was not quite extinguished in his party. They saw that a Union was inevitable, but some, at least, looked beyond it. 'I know,' said Goold, 'that the Ministers must succeed; yet I will not go away with an aching heart, because I know that the liberties of the people must ultimately triumph. The people must at present submit, because they cannot resist 120,000 armed men; but the period will occur when, as in 1782, England may be weak, and Ireland sufficiently strong to recover her lost liberties.' Nor were the last words of Grattan devoid of hope: 'The constitution,' he exclaimed, 'may for a time be lost, but the character of the people cannot be lost. The Ministers of the Crown may perhaps at length find out that it is not so easy to put down for ever an ancient and respectable nation by abilities, however great, or by corruption, however irresistible. Liberty may repair her golden beams, and with redoubled heat animate the country. The cry of loyalty will not long continue against the principles of liberty. Loyalty is a noble, a judicious, and a capacious principle, but in these countries loyalty distinct from liberty is corruption, not loyalty. The cry of the connection will not in the end avail against the principles of liberty. Connection is a wise and a profound policy, but connection without an Irish Parliament is connection without its own principle, without analogy of condition, without the pride of honour that should attend it—is innovation, is peril, is subjugation—not connection. . . . Identification is a solid and imperial maxim, necessary for the preservation of freedom, necessary for that of empire; but without union of hearts, with a separate Government and without a separate Parliament, identification is extinction, is dishonour, is conquest—not identification. Yet I do not give up my country. I see her in a swoon, but she is not dead. Though in her tomb she lies helpless and motionless, still there is on her lips a spirit of life, and on her cheek a glow of beauty:

> Thou art not conquered: beauty's ensign yet
> Is crimson in thy lips and in thy cheeks,
> And death's pale flag is not advanced there.

While a plank of the vessel stands together, I will not leave her. Let the courtier present his flimsy sail, and carry the light bark of his faith with every new breath of wind; I will remain anchored here with fidelity to the fortunes of my country, faithful to her freedom, faithful to her fall.' These were the last words of Grattan in the Irish Parliament.

In England, Sheridan resisted the measure at every step of its progress with persevering earnestness. He moved that its consideration should be delayed till the sentiments of the people of Ireland had been ascertained, but his motion was defeated by 30 to 206. 'I would have fought for that Irish Parliament,' he afterwards exclaimed to Grattan—'ay, up to the knees in blood!' Among the speakers on the measure in the House of Lords was Lord Byron, who described it as the 'union of the shark with its prey.' All opposition, however, was fruitless, and the Bill received the royal assent on August 1.

It has been argued, with much force, and in perfect accordance with the doctrines of the great political writers of the seventeenth century, that the Irish Parliament was constitutionally incompetent to pass the Union. It was the trustee, not the possessor, of the legislative power. It was appointed to legislate, not to transfer legislation—to serve the people for eight years, not to hand over the people to another Legislature. The Act was in principle the same as if the Sovereign of England were to transfer her authority to the sovereign of another nation. It transcended the capacities of Parliament, and was therefore constitutionally a usurpation. 'The Legislature,' in the words of Locke, 'neither must nor can transfer the power of making laws to anybody else, or place it anywhere but where the people have.'

The only qualification which is to be made to this doctrine is a very obvious one. Parliament is the trustee of the nation, but the nation may enlarge its powers, and give it the right to destroy itself. The essential condition of the constitutional validity of an act by which the national representatives destroy the national representation is that the policy of that act should have been submitted to the decision of the constituents. In our own day, no considerable innovation in politics, no material modification in the representative system, could be effected without such an appeal, and it is one of the most important functions of a well-organised House of Lords that it delays contemplated changes until it has been made. But the Union, which swept away a Parliament that had existed for centuries, and had recently been emancipated by the enthusiasm of the entire nation, was carried without a dissolution, without any reference to the voice of the people. It is a memorable fact, indicating the power of the Tory reaction which followed the French Revolution, that when Irish Liberals and English Whig statesmen urged that a question of this kind ought to be brought before the nation by a dissolution, their doctrine was again and again denounced by Pitt as the most palpable and most flagrant Jacobinism. The Government not only showed no desire to consult the wishes of the people, but it even strenuously laboured to separate the representatives from their influence. 'It seems,' wrote the Duke of Portland to Lord Castlereagh in 1799, 'as if the cry of Dublin had carried away many gentlemen whose interests in all respects must be benefited by a Union; and I have seen with some surprise, as well as with real concern, a deference expressed for the opinion of constituents which I conceive to have been as unnecessary as it is entirely unconstitutional.'[1] 'The clamour out of doors,' wrote Lord Cornwallis in the same year, 'is chiefly to be apprehended as furnishing the members within with a plausible pretext for acting in conformity to their own private feelings.'[2]

If the people, however, were not to influence their representatives, there was another kind of influence

---

[1] 'Castlereagh Correspondence,' vol. ii. p. 146.
[2] 'Cornwallis Correspondence,' Jan. 1799.

about which no scruples were entertained. It is, I believe, a simple and unexaggerated statement of the truth, that in the entire history of representative government there is no instance of corruption having been applied on so large a scale, and with so audacious an effrontery, as by the Ministers in Ireland. The trustees of the patronage of the nation, their one object was, by the abuse of their trust, to bribe the representatives to sacrifice their constituents. The constitution of Parliament, in which more than one-third of the seats were nomination boroughs, and a large proportion of these boroughs in four or five hands, gave them fatal facilities, and a long course of adverse influences had made the political classes in Ireland profoundly corrupt. Lord Cornwallis, the Lord-Lieutenant—a brave, frank, and humane soldier, who was sincerely anxious to benefit the empire, and who retained his honourable instincts while discharging a most dishonourable office—felt acutely the task that was confided to him, and in one of his letters applied to himself with remarkable candour the lines of Swift:

> From hell a viceroy devil ascends,
> His budget with corruption crammed,
> The contributions of the damned,
> Which with unsparing hand he strows
> Through courts and Senate as he goes.

'The political jobbery of this country,' he writes, 'gets the better of me. . . . I trust that I shall live to get out of this most cursed of all situations, and most repugnant to my feelings. How I long to kick those whom my public duty obliges me to court!' 'My occupation is now of the most unpleasant nature, negotiating and jobbing with the most corrupt people under heaven. I despise and hate myself every hour for engaging in such dirty work, and am supported only by the reflection that without an Union the British empire must be dissolved.'[1]

Castlereagh, however, who was the more immediate agent in corrupting, appears to have performed his task with a perfect equanimity; and Pitt, the great contriver and organiser of the whole, preserved throughout that tone of lofty piety, and serene, self-complacent virtue, which he knew so well how to assume. All the resources of ecclesiastical, judicial, county, and other patronage were strained to the utmost to find places for those who would support the Union, or to provide for their families and friends; and when these did not suffice, sinecures, pensions, sums from the secret service money, were lavishly employed. A direct, minute system of corruption was applied to every individual whose constancy was not regarded as unassailable. But it was soon found that all this would fail unless measures of a more sweeping kind were taken. The majority of the landlord class, in whose hands the county representation remained, were strongly opposed to the Union; and Castlereagh in 1799 complained bitterly of 'the warmth of the country gentlemen, who spoke in great numbers and with much energy against the question;'[2] but the county seats were immensely outnumbered by the boroughs, and to purchase these was soon found to be necessary. Adopting a plan which he had recommended in 1785 in England, Pitt determined to recognise the patronage of a borough as a form of property, and, in the event of the abolition of the Irish Parliament, to compensate the patrons at the rate of 7,500*l.* a seat. A million and a quarter was expended chiefly in this manner, and the selfish in-

---

[1] 'Cornwallis Correspondence,' vol. iii. pp. 101, 102.
[2] 'Castlereagh Correspondence,' vol. ii. p. 143.

terests of the most influential classes in Ireland passed to the side of the Union, while further compensation was given to other politicians whose interests it was believed would be injuriously affected by the measure.

The precedent of 1800 was afterwards remembered when the English nomination boroughs were abolished in 1832; but all parties indignantly repudiated the notion of recognising such a principle in England. Another mode of corruption scarcely less efficacious was employed to influence the wealthier Irish gentry. Peerages to this class are always a peculiar object of ambition, and they had long been given in Ireland with a lavishness which materially degraded the position. In England, the simultaneous creation of twelve peers by Harley had been regarded as a scandalous and unprecedented straining of the prerogative; but no sooner had the Union been carried than Lord Cornwallis sent to England the names of sixteen persons, to whom he had expressly promised Irish peerages as rewards for their support of the Union. But these promotions were but a small part of what was found necessary; twenty-two Irish peers were created, five peers received English peerages, and twenty peers received higher titles.

It was a boast of Lord Castlereagh that, whatever might be the case with the majority of the Irish people, the preponderance of landed property was unquestionably on the side of the supporters of the Union. In making this calculation, he took into account the Irish peers, who were chiefly subservient to the Government; the bishops, who were very wealthy, and who, with two exceptions, voted for the Union; and the great English noblemen who possessed estates in Ireland: but he also maintained that the balance of property in the Commons, though not in the same degree, was on the same side.[1] Considering the part that was taken by the county members, this last calculation seems very questionable, but, if it be true, it is not difficult to explain it. The Ministers, by money or by dignities, had bought almost all the great borough-owners, as well as a large proportion of the members, and they thus made their success certain. One difficulty, however, still remained. It was found that several of the borough members were not prepared to vote for the Union, although their patrons had been bought. The most obvious way of meeting this difficulty would have been to have dissolved Parliament, but such a step would have given the free constituencies an opportunity of testifying their abhorrence of the measure. A simpler method was accordingly adopted. A place Bill, intended to guard the purity of Parliament against the corruption of Ministers, by compelling all who accepted office to vacate their seats, had been recently passed, and the Ministers ingeniously availed themselves of this to consummate the triumph of corruption. According to the code of honour which then prevailed both in England and Ireland, the members of nomination boroughs who were unwilling to vote as their patrons directed considered themselves bound to accept nominal offices, and thus vacate their seats, which were at once filled by staunch Unionists, in some instances by English and Scotch men wholly unconnected with Ireland.

By these means the majority was formed which sold the constitution of Ireland. Lord Cornwallis, in a private letter, described its character with perfect frankness. 'The nearer the great event approaches,' he wrote, 'the more are the needy and interested senators alarmed at the effects it may possibly have on

[1] 'Cornwallis Correspondence,' vol. iii. p. 224.

their interests and the provision for their families, and I believe that *half of our majority* would be at heart as much delighted as any of our opponents if the measure could be defeated.'[1] Grattan, who had unusual opportunities of judging, afterwards expressed his opinion that, of the members who voted for the Union, only seven were unbribed.[2]

While these events were taking place, the nation, as I have said, was prostrated and exhausted by the Rebellion. A fierce animosity divided the Catholics and Protestants; the country was full of English troops; and the reign of martial law, as well as the reaction of exaggerated loyalty that always follows insurrection, made men more than commonly timid in opposing the Government. A considerable proportion of the Catholic priests had been successfully bribed by the hope of payment, commutation of tithes, and emancipation. Their flocks, through fear, influence, or resentment, were chiefly passive, and the wealthiest Protestant proprietors had been purchased by peerages or places. But, notwithstanding all this, twenty-eight counties, twenty of them being unanimous, petitioned against the Union; and the petitions against it are said to have had more than 700,000 signatures, while those in its favour had only 7,000.[3] Of the Irish Parliament itself, Mr. Grey, in the English House of Commons, gave the following analysis: 'There are 300 members in all, and 120 of these strenuously opposed the measure, among whom were two-thirds of the county members, the representatives of the city of Dublin, and almost all the towns which it is proposed shall send members to the Imperial Parliament: 162 voted in favour of the Union; of

---

[1] 'Cornwallis Correspondence,' vol. iii. p. 228.
[2] Grattan's Life, vol. v. p. 113.   [3] Ibid. p. 51.

these, 116 were placemen. Some of them were English generals on the staff, without a foot of ground in Ireland, and completely dependent upon Government.'[1] In the case of Ireland, as truly as in the case of Poland, a national constitution was destroyed by a foreign Power, contrary to the wishes of the people. In the one case, the deed was a crime of violence; in the other, it was a crime of treachery and corruption. In both cases a legacy of enduring bitterness was the result.

There are, indeed, few things more discreditable to English political literature than the tone of palliation, or even of eulogy, that is usually adopted towards the authors of this transaction. Scarcely any element or aggravation of political immorality was wanting, and the term honour, if it be applied to such men as Castlereagh or Pitt, ceases to have any real meaning in politics. Whatever may be thought of the abstract merits of the arrangement, the Union, as it was carried, was a crime of the deepest turpitude—a crime which, by imposing, with every circumstance of infamy, a new form of government on a reluctant and protesting nation, has vitiated the whole course of Irish opinion. The loyalty of a nation is chiefly due to the associations formed by the events of its history, but these in Ireland have tended with a melancholy uniformity in the opposite direction. I have already observed how the three greatest English rulers, Elizabeth, Cromwell, and William III., are associated in Ireland with memories of disaster and humiliation, and how the prostration of English power in America produced the Irish declaration of independence. Flood desired to employ the Volunteer movement to coerce the Parliament and Government into a reform, but the policy of Grattan and Charlemont prevailed. The

[1] Hansard, vol. xxxv. p. 60 (April 21, 1800).

Volunteers, with a signal loyalty, disbanded, and left the question of reform to the constitutional forces of the nation. The result was that the prediction of Flood was fully verified; the corruption of Parliament was carefully maintained and aggravated, and it was ultimately made the means of destroying the constitution. The danger that menaced England from France in 1793 produced the concession of the suffrage to the Catholics. The weakness of Ireland after the Rebellion was made an opportunity of depriving her of her Legislature. The prospect of emancipation and of the commutation of tithes was held out to the Catholics, and they, for the most part, abstained in consequence from actively opposing the Union, but when the measure had been carried Pitt sacrificed them with little more than a show of reluctance to the King. Eventually, however, the Catholics were emancipated and the tithes commuted, but the first measure was due to an agitation which brought the country to the verge of civil war, and the second was the reward of almost universal rebellion against the law. Thus, generation after generation, by a slow, steady, and fatal process, the nation has been educated into disloyalty, taught to look with distrust upon constitutional means of obtaining its ends, and accustomed to regard outrage and violence as the invariable preludes of concession.

That the Parliament which was swept away in 1800, and the political classes it represented, were exceedingly corrupt cannot reasonably be questioned. Almost all the honest patriots of the country were, I believe, on the side of the Opposition, and there were men among them who would have done honour to any legislative assembly; but it is a mere delusion to regard the opponents of the Union as exclusively or mainly actuated by pure patriotism. Selfish local and personal motives contributed largely to their opposition, and they also attempted to carry their ends by corruption. When, however, the undoubted venality of the Parliament is urged as an apology for the Union, an Irish writer may be permitted to remind his readers to whom that venality is to be ascribed—who resisted every serious effort of reform. The corruption of the Legislature had been made a main function of the Government, and it was successfully accomplished. If, however, the spectacle presented by the majority in that Legislature in 1800 was eminently despicable, it should not be forgotten that there is no other instance in history of such extensive corruption being applied to a legislative body; that in the first year, when the Union was brought forward, Parliament was proof against temptation; that the measure was ultimately carried by introducing into the nomination boroughs new members, in some instances wholly unconnected with Ireland; and that, defective as the constitution of Parliament undoubtedly was, it is extremely questionable whether the Union could have been carried had there been a dissolution. It must be added, too, that the corruption of the House of Commons was not so great as to prevent it on important occasions from yielding to the wishes of the people. The Irish House of Lords was a perfectly subservient body; the Irish House of Commons never was. In the early part of the eighteenth century the refractory element in it was chiefly due to the extreme dislike of the Irish landlords to tithes, while the English interest was for a long space of time directed by the Primates of the Church. Archbishop Boulter complains bitterly of the opposition he had on this ground to encounter, and of the desire of the Parliament on every occasion to injure the Church.[1] At a later period the Octennial Bill was forced by public opinion on a very reluctant Parliament, and Parliament fully reflected the national enthusiasm in 1782. In Ireland, as in England, a certain proportion of the borough-owners were patriotic, and several of them came forward prominently in support of the Reform Bill of Flood.[2] Parliamentary reform in Ireland would undoubtedly have been very difficult, but, had the Parliament continued, it would at last have been effected, as in England, by the influence of the Government, aided by the defection of some of the borough-owners, and supported by an overwhelming pressure of public opinion. The Irish Parliament, though very corrupt if compared with the British Parliament at the end of the last century, and of course still more with that of our own day, was probably not much more corrupt, and was certainly much more tolerant, than that which sat in London in the early years of the eighteenth century. It was guided habitually by sordid motives, but it not unfrequently rose above them; and this is about as much as can be said for not a few of the Parliaments of England. No one has stigmatised the Irish Legislature in more vehement terms than Lord Macaulay, but he could hardly apply to it stronger terms of condemnation than he applied to the English Parliament of Walpole, 'who governed by corruption, because in his time it was impossible to govern otherwise.' 'A large proportion of the members,' we are told, 'had absolutely no motive to support any Administration except their own interest, in the lowest sense of the word. Under these circumstances, the country could be governed only by corruption. . . . We might as well accuse the poor Lowland farmers,

[1] See Boulter's Letters, vol. ii. pp. 154, 217, 234–236.
[2] Grattan's Life, vol. iii. p. 123.

who paid blackmail to Rob Roy, of corrupting the virtue of the Highlanders, as accuse Sir R. Walpole of corrupting the virtue of Parliament.'

But, in truth, the clouds of enthusiasm or obloquy which during the Repeal contest gathered so thickly around this portion of Irish history, make it even now difficult for either English or Irish writers to pronounce with perfect impartiality on the merits of the old Parliament of Ireland. The time may come when the historians of other nations may review its history, and I cannot but think that, while they will find much to blame, they will find in its later years at least something to admire. In estimating the character of a Legislature, we should consider the period of its existence, the difficulties with which it had to contend, and the temptations to which it was exposed ; and if these things be taken into account, the Irish Parliament will not be wholly condemned. Seldom has even the Imperial Parliament exhibited a constellation of genius more brilliant, more varied, and more pure than that which is suggested by the names of Grattan and Flood, of Curran, Plunket, Hutchinson, and Burrowes. That a Legislature so defective in its constitution should have continued to exist is indeed wonderful, but it is far more wonderful that it should have achieved what it did—that it should have asserted its own independence—that it should have riven the chains that fettered its trade—that it should have removed the most serious disabilities under which the mass of the people laboured—that it should have voluntarily given up the monopoly of power it possessed, as representing the Protestants alone. With every inducement to religious bigotry, it carried the policy of toleration in many respects much farther than the Parliament of England. With every inducement to disloyalty, it was steadily faithful to the connection. And its reputation has suffered by its fidelity, for the bitter invectives of the United Irishman Wolfe Tone have been reproduced by English writers as if they were the most impartial description of its merits.[1]

'I argue not,' said Grattan, 'like the Minister, from the misconduct of one Parliament against the being of Parliament itself. I value that Parliamentary constitution by the average of its benefits, and I affirm that the blessings procured by the Irish Parliament in the last twenty years are greater than the blessings afforded by British Parliaments to Ireland for the last century; greater even than the mischiefs inflicted on Ireland by British Parliaments; greater than all the blessings procured by these Parliaments for their own country within that period. Within that time the Legislature of England lost an empire, and the Legislature of Ireland recovered a constitution.'

Nor should it be omitted that the Irish Parliament was on the whole a vigilant and intelligent guardian of the material interests of the country. During the greater part of the century, indeed, it had little power except that of protesting against laws crushing Irish commerce; but what little it could do it appears to have done. Its Journals show a minute attention to industrial questions, to the improvement of means of communication, to the execution of public works. One of the most important events in English industrial history in the eighteenth century was the creation of a system of inland navigation by means of canals with locks—an

---

[1] Thus, *e.g.*, Macaulay, in his very fine speech 'on the state of Ireland,' having poured a multitude of fierce epithets on the Irish Parliament, concluded with this very singular sentence: 'I do not think that by saying this I have given offence to any gentleman from Ireland, however zealous for repeal he may be, for I *only* repeat the language of Wolfe Tone.'

improvement which is due to the genius of the engineer Brindley, and to the enterprise of the Duke of Bridgewater. The first canal of any magnitude in England was that between Worsley and Manchester, which was opened in 1761. The experiment was practically a new one, for, with one very inconsiderable exception, there was no other canal in England. But the Irish Parliament appears to have immediately perceived the importance of the enterprise, and the energy and alacrity with which it undertook to provide Ireland with a complete system of internal navigation is beyond all praise. In 1761 it voted a sum of 13,500*l*. to the corporations of several inland navigations, and made special grants for a canal from Dublin to the Shannon, and for improving the navigation of the Shannon, Barrow, and Boyne. Two years later works of the most extensive kind appear to have been undertaken. Among the votes of the Irish House of Commons in 1763 we find grants for the construction of a canal between Dublin and the Shannon; for a canal from Newry to Lough Neagh; for a canal connecting Loch Swilly and Loch Foyle; for a canal which, together with improvements on the river Lagan, was intended to complete the navigation between Loch Neagh and the sea at Belfast; and for four other inland navigations by canals.[1] In the last years of the Irish Parliament, or at all events from the concession of free trade in 1779 to the Rebellion of 1798, the material progress of Ireland was rapid and uninterrupted. In ten years, from 1782, the exports more than trebled.[2] Lord Sheffield, who wrote upon Irish commerce in 1785, said, 'At present, perhaps, the improvement of Ireland is as rapid as any country ever experienced;' and Lord

[1] Macpherson's 'Annals of Commerce,' vol. iii. pp. 349-383.
[2] See Grattan's Speech, May 18, 1810.

Clare, in a pamphlet which appeared in 1798, made a similar assertion with much greater emphasis. Speaking of the period that had elapsed since 1782, he said, 'There is not a nation in the habitable globe which has advanced in cultivation and commerce, in agriculture and manufactures, with the same rapidity in the same period.'

The dangers to the connection which have been supposed to spring from the existence of the Irish Parliament have been chiefly illustrated by the Rebellion of 1798, and by the dissension on the Regency question. The former may be very rapidly disposed of; for to identify it in any degree with the independence of Parliament is to manifest a complete ignorance of the facts of the case. The Rebellion of '98 was produced by exceptional causes—by the excitement consequent upon the French Revolution, acting upon the excitement consequent on the recall of Lord Fitzwilliam. It was not represented by any party in Parliament. Grattan, who was the leader of the Irish Whigs, was so bitterly opposed to everything French that he completely separated himself on French questions from Fox and the English Whigs, with whom he generally acted, and who looked with favour on the Revolution. He once went so far as to speak of 'eternal friendship with France' as one of the 'curses' to which Ireland would be doomed if emancipation were withheld. On the other hand, perhaps, no one ever hated the Parliament more than the United Irishmen. The people rebelled, not because there was an organ of public opinion in the land, but because that organ, while unreformed, did not sufficiently represent the national feeling. It was the energetic exertion of the Parliament that repressed the Rebellion before the arrival of the English troops; and had it not been for its prompt and decisive action, it is difficult to say how far the movement might have spread.

The difference which arose between the English and Irish Parliaments concerning the Regency was undoubtedly a very serious embarrassment; but its constitutional importance has, I think, been greatly exaggerated. It was admitted on all sides that the Sovereign possessed the same plenitude of power in Ireland as in England, but the question which arose when he had been incapacitated by insanity was absolutely novel and unprecedented. It had been foreseen by no statesman, and nothing in past English history was of any real assistance in solving it. The English Parliament decided that the authority of providing for the discharge of the functions of royalty reverted to Parliament, which had a right to impose what restrictions it pleased upon the Regent. The Irish Parliament, adopting, it may be observed, the more modest view of the functions of Parliament—a view which has recently been defended by the high authority of Lord Campbell —maintained that in an hereditary monarchy the eldest son of the Sovereign has the same absolute right to his father's place during the incapacity, as he would have after the death, of the latter. The difference was no doubt perplexing, and for a time dangerous; but it was extremely easy to guard against the possibility of its recurrence by a special law providing that whoever exercised the power of Regent *de facto* in England should exercise a similar power *de jure* in Ireland. A corresponding legal maxim was already recognised in the case of the Sovereign; there would have been no real difficulty in extending it to the case of the Regent, and a resolution to that effect was actually brought forward by the anti-Union party.

If, however, we consider the question in a more

general point of view, it must be admitted that a collision between two independent Legislatures was by no means an unlikely event; and it is impossible to doubt that in that case the connection might be seriously endangered. The peril from this source was real and grave; and it appears to me plain that for this, as for other reasons, the system of 1782 must eventually have been modified. At the same time the danger has been overrated; and were it otherwise, a premature Union unaccompanied by emancipation was not the proper way of averting it. A very similar danger exists in the British constitution itself, for if a difference arose between its three constituent elements, in which each obstinately refused to yield, Government might be brought to a dead lock, or the nation to a revolution or a war of classes. The complexity of the constitution is retained, not because such a catastrophe is impossible, but because it is believed that the advantages preponderate over the disadvantages—because, although under certain circumstances that complexity might create discord and revolution, it is on the whole admirably calculated to prevent or allay them. The blended force of interest and patriotism inspire the Sovereign, the aristocracy, and the Commons with the spirit of compromise, which is essential to their co-operation. It is not true that independent Legislatures cannot be so constituted or their limits of action so defined that they should work in harmony. The Colonial Legislatures in the British empire are a striking proof to the contrary, and the federal principle which has existed for ages in the only flourishing European Republic, and which has contributed so largely to the wellbeing of the great Republic of the West, has of late years been advancing with considerable strides through monarchical Europe. At any period of the eighteenth century

England might easily have bound the Irish Legislature to herself by ties of interest of overwhelming force; for by the concession of free trade, and by throwing open to Irishmen the great careers of colonial administration, she could have made the connection a matter of vital importance to Ireland. That it is possible for reckless or ignorant agitators to disregard such considerations of national interest is but too true; but it is hardly possible that they could fail to exercise a restraining influence upon a Parliament, or a public opinion, which was guided by the property and the intelligence of the country.

But in truth the harmonious co-operation of Ireland with England depends much less upon the framework of the institutions of the former country than upon the dispositions of its people and upon the classes who guide its political life. With a warm and loyal attachment to the connection pervading the nation, the largest amount of self-government might be safely conceded, and the most defective political arrangements might prove innocuous. This is the true cement of nations, and no change, however plausible in theory, can be really advantageous which contributes to diminish it. Theorists may argue that it would be better for Ireland to become in every respect a mere province of England; they may contend that a union of Legislatures, accompanied by a corresponding fusion of characters and identification of hopes, interests, and desires, would strengthen the empire, but as a matter of fact this was not what was effected in 1800. The measure of Pitt centralised, but it did not unite, or rather by uniting the Legislatures it divided the nations. In a country where the sentiment of nationality was as intense as in any part of Europe, it destroyed the national Legislature contrary to the manifest wish of the people, and by means so corrupt, treacherous, and shameful that they are never likely to be forgotten. In a country where, owing to the religious differences, it was peculiarly necessary that a vigorous lay public opinion should be fostered to dilute or restrain the sectarian spirit, it suppressed the centre and organ of political life, directed the energies of the community into the channels of sectarianism, drove its humours inwards,[1] and thus began a perversion of public opinion which has almost destroyed the elements of political progress. In a country where the people have always been singularly destitute of self-reliance, and at the same time eminently faithful to their leaders, it withdrew the guidance of affairs from the hands of the resident gentry, and, by breaking their power, prepared the ascendency of the demagogue or the rebel. In two plain ways it was dangerous to the connection: it incalculably increased the aggregate disloyalty of the people, and it destroyed the political supremacy of the class that is most attached to the connection. The Irish Parliament, with all its faults, was an eminently loyal body. The Irish people through the eighteenth century, in spite of great provocations, were on the whole a loyal people till the recall of Lord Fitzwilliam, and even then a few very moderate measures of reform might have reclaimed them. Burke, in his 'Letters on a Regicide Peace,' when reviewing the elements of strength on which England could confide in her struggle with revolutionary France, placed in the very first rank the co-operation of Ireland. At the present day it is to be feared that most impartial men would

---

[1] 'To give moderate liberty for griefs and discontentments to evaporate ... is a safe way; for he that turneth the humours back, and maketh the wound bleed inwards, endangereth malign ulcers and pernicious imposthumations'—*Bacon, On Seditions.*

regard Ireland in the event of a great European war rather as a source of weakness than of strength. More than seventy years have passed since the boasted measure of Pitt, and it is unfortunately incontestable that the lower orders in Ireland are as hostile to the system of government under which they live as the Hungarian people have ever been to Austrian or the Roman people to Papal rule; that Irish disloyalty is multiplying enemies of England wherever the English tongue is spoken; and that the national sentiment runs so strongly that multitudes of Irish Catholics look back with deep affection to the Irish Parliament, although no Catholic could sit within its walls, and although it was only during the last seven years of its independent existence that Catholics could vote for its members. Among the opponents of the Union were many of the most loyal as well as nearly all the ablest men in Ireland; and Lord Charlemont, who died shortly before the measure was consummated, summed up the feelings of many in the emphatic sentence with which he protested against it. 'It would more than any other measure,' he said, 'contribute to the separation of two countries, the perpetual connection of which is one of the warmest wishes of my heart.'

In fact the Union of 1800 was not only a great crime, but was also, like most crimes, a great blunder. The manner in which it was carried was not only morally scandalous; it also entirely vitiated it as a work of statesmanship. No great political measure can be rationally judged merely upon its abstract merits, and without considering the character and the wishes of the people for whom it is intended. It is now idle to discuss what might have been the effect of a Union if it had been carried before 1782, when the Parliament was still unemancipated; if it had been the result of a spontaneous movement of public opinion; if it had been accompanied by the emancipation of the Catholics. Carried as it was prematurely, in defiance of the national sentiment of the people and of the protests of the unbribed talent of the country, it has deranged the whole course of political development, driven a large proportion of the people into sullen disloyalty, and almost destroyed healthy public opinion. In comparing the abundance of political talent in Ireland during the last century with the striking absence of it at present, something no doubt may be attributed to the absence of protection for literary property in Ireland in the former period, which may have directed an unusual proportion of the national talent to politics, and something to the Colonial and Indian careers which have of late years been thrown open to competition; but when all due allowance has been made for these, the contrast is sufficiently impressive. Few impartial men can doubt that the tone of political life and the standard of political talent have been lowered, while sectarian animosity has been greatly increased, and the extent to which Fenian principles have permeated the people is a melancholy comment upon the prophecies that the Union would put an end to disloyalty in Ireland.

While, however, the Irish policy of Pitt appears to me to be both morally and politically deserving of almost unmitigated condemnation, I cannot agree with those who believe that the arrangement of 1782 could have been permanent. The Irish Parliament would doubtless have been in time reformed, but it would have soon found its situation intolerable. Imperial policy must necessarily have been settled by the Imperial Parliament in which Ireland had no voice; and, unlike Canada or Australia, Ireland is profoundly affected by every change of Imperial policy. Connection with England was of overwhelming importance to the lesser country, while the tie uniting them would have been found degrading by one nation and inconvenient to the other. Under such circumstances a Union of some kind was inevitable. It was simply a question of time, and it must some day have been demanded by Irish opinion. At the same time it would not, I think, have been such a Union as that of 1800. The conditions of Irish and English politics are so extremely different, and the reasons for preserving in Ireland a local centre of political life are so powerful, that it is probable a federal Union would have been preferred. Under such a system the Irish Parliament would have continued to exist, but would have been restricted to purely local subjects, while an Imperial Parliament, in which Irish representatives sat, would have directed the policy of the empire.

It remains only to add a few words upon the manner in which the Ministers observed their pledges to the Catholics. After the deadly injury which had been done them by the recall of Lord Fitzwilliam, it might have been supposed that a statesman of common uprightness would have been peculiarly anxious that they should have no further ground of complaint. Lord Cornwallis and Lord Castlereagh, as the representatives of the Government, had purchased the support of the leading Catholic prelates by a distinct intimation that in their opinion the Union would be a prelude to emancipation. Without giving any express promise which could impede the Union negotiations, they had excited their hopes by assuring them that the Ministers were sincerely and unanimously in favour of the principle of emancipation, and on the faith of that assurance they had solicited and obtained a most important service.

The great body of the Catholics had been induced to remain passive; and if the Catholics had been in active opposition, the Union, in the opinion of Lord Castlereagh, could not have been carried. Whatever might be the exact terms of the intimation made to the Catholic leaders, no statesman with a high sense of honour could question that the Cabinet were bound to do the very utmost in their power to carry emancipation. It was an obligation of honour of the plainest kind, and it was also a matter of policy of the most vital importance. The Union, carried as it was, outraged every patriotic and national sentiment in the country; and if it was not to be a source of the most perennial bitterness, it was absolutely necessary that it should be accompanied or speedily followed by some great national boon, which might at least make some class of Irishmen look back on it with satisfaction. The Scotch Union had thrown open to Scotchmen the whole trade with the English Colonies in America, from which they had before been excluded, but this trade had been thrown open to Irishmen in 1779. Free trade between England and Ireland was indeed established by the Union; but this advantage, though a very important one, was not sufficiently great or sufficiently calculated to strike the imagination to counteract its evil effects. Catholic emancipation alone could have the required effect, and on the conduct of the Ministers at this momentous juncture it depended whether the Catholics were to be permanently loyal. Duped and injured as they had been in 1795, their loyalty was not likely to bear the strain of a second disappointment.

It seemed at first as though the Government would do everything that could be expected. In the first King's Speech after the Union, the Sovereign was made to describe it as the happiest event of his reign;

'being persuaded,' as the Speech continued, 'that nothing could so effectually contribute to extend to my Irish subjects the full participation of the blessings derived from the British constitution.' It is not very clear what meaning these expressions conveyed to the Sovereign who used them; but the Catholic leaders naturally read them in the light of the negotiations that had taken place, and as naturally interpreted them as a promise of emancipation. They assumed that the Catholics, who constituted three-quarters of the Irish people, were included under the denomination of 'Irish subjects,' and that the right of sitting in Parliament was one of the blessings of the constitution. It soon, however, appeared that the King was vehemently opposed to emancipation; and the Chancellor, Lord Loughborough, through selfish, and the Primates of England and Ireland through ecclesiastical motives, inflamed his opposition. While his Ministers were bribing the Catholics to acquiesce in the Union by holding out to them the hope that it would secure their emancipation, the King was basing his policy on a directly opposite calculation. 'My inclination to the Union with Ireland,' he wrote in February 1801, 'was chiefly founded on a trust that the uniting of the Established Churches of the two kingdoms would for ever shut the door to any further measures with respect to the Roman Catholics.' The language which had been held to the Catholics, and in reliance on which they had in general abstained from opposing the Union, had been held without the knowledge of the King, and without the smallest attempt having been made to learn how far his antipathy might be surmounted. This was in itself sufficiently culpable; but after all that had been said and done, it is at least plain that Pitt was under the strongest moral obligation to do the utmost in his power to carry the measure. The King talked of abdicating if it were passed; but even that alternative should have been faced, though it should not be forgotten that the King was accustomed to use such threats whenever he urgently desired to carry his point, and that his language about the recognition of the independence of America, and about the admission of Fox into his Cabinet, was quite as strong as his language about Catholic emancipation. It was an imperious obligation of national honour—it was a matter of vital importance to the future prosperity of the empire—that the Catholics should at this time have been emancipated, and there is no reasonable doubt that Pitt could have carried the measure had he determined it.

He did, it is true, resign office when the King refused to consent to it; but there has seldom been a resignation which deserves less credit. The step was evidently taken solely because it was impossible that he could have acted otherwise with any decorum or without a palpable loss of character, and because Lord Grenville and some of his other colleagues had a strong and honourable sense of their duty to the Catholics. It is, however, quite plain that Pitt, having obtained the service he required from the Catholics, felt no real interest in their emancipation; that he was resolved to incur for their sakes no difficulty he could possibly avoid, and was ready, on the first decent pretext, to sacrifice them. He had no personal objection to Catholic emancipation; and on this, as on most other subjects, his views were large and liberal; but on this, as on most other questions, he showed himself thoroughly selfish and dishonest, prepared to sacrifice any principle or any class rather than imperil his power or weaken or divide his followers.

He resigned office into the hands of Addington, whom he regarded as a mere creature of his own, and from whom he imagined he might at any time resume it. He resigned it at a moment which was peculiarly convenient to him, because it had become necessary to negotiate with Napoleon, and the antecedents of Pitt rendered such a negotiation more difficult and humiliating for him than for any other English statesman.[1] He resigned it with his usual ostentatious display of public principle, because the King would not consent to Catholic emancipation; but when the transfer of office had been effected, and when the agitation produced by the transaction threw the King into one of his many attacks of temporary insanity, Pitt immediately availed himself of the opportunity to extricate himself from a political embarrassment by finally abandoning the Catholics. That his position, in consequence of the King's attack, was a delicate one, may be readily admitted; but there was a question of honour and a question of national policy which should have overridden all other considerations; and he would have deserved more credit for his delicacy if it had not coincided so perfectly with his interest, and if it had not involved him in what may be not unfairly called a gross breach of faith with the Catholics. And, in fact, the utmost the most

---

[1] I have no doubt that the Catholic question was the real as well as the ostensible cause of the resignation, but the consideration in the text was an obvious one, and it greatly mitigated the sacrifice. Dundas said of Addington, 'If these new Ministers stay in and make peace, it will only smooth matters the more for us afterwards;' and Lord Malmesbury, who records this saying in his Diary, mentions the impression that 'Pitt is inclined to let this Ministry remain in office long enough to make peace, and then turn them out.'—See Campbell's 'Chancellors,' viii. pp. 190, 191; and a remarkable letter by Dean Milman in Lewis's 'Administrations of Great Britain,' p. 270.

sensitive delicacy required was that he should have abstained at the time of the King's illness from pressing the question. But this was not the course which he adopted. Ostensibly through attachment to the cause of Catholic emancipation, he resigned his office into the hands of a violent anti-Catholic statesman, who, as we now know, assumed it at his express desire. Only three weeks later, when the King had recovered, when Addington had formed his Ministry without difficulty, and when all was proceeding smoothly, he volunteered the announcement to the King that he would never during the King's life bring forward the Catholic question; and he desired by this means, if the King or Addington would take the first step, to return to power. This was the end of the 'unalterable sense of public duty' which had led him, as he declared three weeks before, to resign office because he was not allowed to bring in the Catholic question with his Majesty's 'full concurrence, and with the whole weight of Government.' This was the end of all the hopes by which Castlereagh had lulled to sleep the Catholic opposition to the Union. Addington, it is true, refused to be treated as a mere puppet, and to resign the dignity he had just been entreated to assume; but the treachery of Pitt was only postponed. He soon became Minister again, and he resumed the reins of power on the understanding that he would not only not bring in Catholic emancipation during the King's lifetime, but that he would also not suffer it to be carried. As for the payment of the priests, which was another important part of the Union scheme, he never appears to have taken any real trouble on the subject.

In the meantime, great apprehension was felt about the attitude of the Irish Catholics. Except during the brief interval of tranquillity which followed the peace of Amiens, England was engaged in the most desperate struggle with France, and Catholic disloyalty appeared proportionately terrible. Immediately upon the resignation of Pitt, and the installation of a new and anti-Catholic Ministry, the Lord-Lieutenant, Lord Cornwallis, drew up a paper, which he privately circulated among the Catholic leaders, in which he earnestly exhorted them to patience under their disappointment, warned them against Jacobinical associations, and expatiated upon the great advantage their cause had gained in so many eminent statesmen being pledged not to take office without carrying it. This paper was unofficial, but, emanating as it did from the Lord-Lieutenant, it had naturally great weight. It proved however to be but one more added to the many deceptions the Irish Catholics experienced. Lord Cornwallis, who immediately after resigned his office, subsequently admitted that he had no authority for the statement that the retiring Ministers were pledged to abstain from office till they could carry Catholic emancipation. He had merely drawn an inference—though it must be admitted a very natural inference—from the situation. Whatever may have been the opinion of others, *he* at least believed that the communications he had made to the Catholic leaders amounted to a moral pledge. When Pitt, three weeks after his resignation, offered to abandon the Catholics, he made none of his colleagues his confidants except Dundas, who was notorious among politicians for his lax sense of honour; but on his return to office, the attitude he resolved to assume towards them became manifest. They acted with the most signal moderation. They would at this time have gladly accepted emancipation accompanied by those safeguards which a few years later they so scornfully rejected. They abstained, not only from all disloyal associations, but even from all political agitation that might embarrass the Government; and it was only in 1805 that their leaders brought over to London a petition for emancipation, which they asked Pitt, who was then in power, to present and to support. He not only refused to do so, but even declared that he would oppose it; and, after a brilliant debate, the Catholics were defeated by an overwhelming majority through his influence. Can it be wondered that O'Connell found them apt scholars when he taught them to exchange a policy of moderation for one of violent agitation?

Grattan, in one of his speeches, described a portion of the English policy towards Ireland with characteristic energy, as one 'than which you would hardly find a worse if you went to hell for your principles, and to Bedlam for your discretion.' I shall content myself with saying that we should have heard few eulogies of the honourable character of the Irish policy of Pitt if English writers were not accustomed to judge Irish politics by a standard of honour very different from that which they would apply to English ones. How his desertion of the Catholics was regarded by the most upright of his opponents is abundantly shown in the private letters of Fox and of Grey; and the subsequent career of O'Connell is a sufficient comment upon the wisdom of his proceedings. It has been maintained, however, by some writers, who would probably have admitted that in these negotiations the part played by Pitt was very culpable, that the original scheme of the Union was at least an extraordinary instance of political genius. Lord Macaulay, who has probably done more than any other writer to accredit this opinion, has described the project of combining in a single measure the legislative Union of the two countries, the emancipation of the Catholics, and the payment of their priests, as 'a scheme of policy so grand and so simple, so righteous and so humane, that it would alone entitle him to a high place among statesmen.' I venture to think that this judgment is entirely erroneous. The project of a Union, and the project of settling the Catholic question by admitting Catholics to Parliament, and by paying their priests, were no novelties. They had for years been commonplace subjects of discussion in political circles; and one of the standard arguments against emancipating the Catholics had been that it would be dangerous to give them such power in a local Parliament. The expediency of combining the two projects was perfectly obvious. The idea was so self-evident that it must have been suggested at a hundred dinner-tables, and it is hardly conceivable that it should not have occurred to any statesman who approved of both measures, and who was seeking to make the first popular in Ireland. The Union was emphatically one of that class of measures in which the scope for statesmanship lies not in the conception but in the execution. Had Pitt carried it without offending the national sentiment—had he enabled the majority of the Irish people to look back on it with affection or with pride—had he made it the means of allaying discontent or promoting loyalty—he would indeed have achieved a feat of consummate statesmanship. But in all these respects he utterly failed. There was, it is true, no small amount of dexterity of a somewhat vulpine order displayed in carrying the Bill; but no measure ever showed less of that enlightened and far-seeing statesmanship which respects the prejudices and conciliates the affections of a nation, and thus eradicates the seeds of disaffection and discontent.

When the Union was passed, Grattan for a time retired from politics. His health had been for some time unsatisfactory, and his spirits were greatly depressed by a defeat which he regarded as the destruction of the liberties of his country. He saw in it the overthrow of the entire labour of his life, and it unfolded to his piercing eye a long vista of agitation, of disloyalty, and disaster. For some time he could not bear to hear it discussed in conversation; his eyes often filled with tears when speaking of it, and even many years afterwards he occasionally broke into paroxysms of indignation on the subject, that contrasted strangely with his usual gentleness.[1] The people, who had been paralysed by the late Rebellion, remained in a state of stupefied and sullen quiescence. Emmett's rebellion, which took place in 1803, cannot be regarded as in any degree the consequence of the Union. It was but the last wave of the Rebellion of 1798, and originated in the overheated brain of an amiable and gifted, but most unpractical, enthusiast. One great cause, however, still remained, and to the service of Catholics Grattan resolved to devote his remaining years. He entered the British Parliament in 1805, and took his seat modestly on one of the back benches; but Fox, exclaiming 'This is no place for the Irish Demosthenes!' drew him forward, and placed him near himself. Great

---

[1] He believed that the Union, among other effects, would have that of greatly lowering the character of the Irish representatives, and he expressed his opinion with his usual odd emphatic exaggeration. 'You have swept away our constitution,' he once said to some English gentlemen; 'you have destroyed our Parliament, but we shall have our revenge. We will send into the ranks of *your* Parliament, and into the very heart of *your* constitution, a hundred of the greatest scoundrels in the kingdom!'

doubts were felt about his success. The difference of the tone and habits of the two Parliaments, the advanced age of Grattan, the recent failure of Flood, and the cause Grattan had assigned for that failure,[1] suggested weighty reason for fear. Much anxiety, therefore, and much curiosity, were felt when he rose to speak on that memorable night when the Catholic question was re-opened. For a moment, it is said, the strangeness of his gestures, and the apparent difficulty of his enunciation, served to confirm those fears; but it was but for a moment. After almost the first passage he was listened to with an intense and ever-increasing admiration, and when he sat down it was felt that he had more than justified his reputation. It was, indeed, one of the very greatest speeches he ever delivered. It would be difficult to point out any other that displayed a more wonderful combination of powerful reasoning, epigram, imagination, and declamation. Pitt, who made the first motion of applause, exclaimed, 'Burke told me that Grattan was a wonderful man for a popular audience, and I see that he was right.' Fox, in a private letter to Trotter, said, 'I am sure it will give you pleasure to hear that Grattan's success in the House of Commons was complete, and acknowledged even by those who had entertained great hopes of his failure.' The 'Annual Register' called the speech 'one of the most brilliant and eloquent ever pronounced within the walls of Parliament.' It was in the course of this speech that, in adverting to the first Catholic Relief Bill, he digressed into an eulogium of the Irish Parliament; and, speaking of the services he had rendered to its freedom, uttered that sentence so famous for its

---

[1] 'He was an oak of the forest too old and too great to be transplanted at fifty.'

touching and concentrated beauty: 'I watched by its cradle, I followed its hearse.'

The Union, by making the public opinion of England the arbiter of the Catholic question, had entirely altered its conditions; and, as I have already endeavoured to show, had considerably increased its difficulties. Public opinion had also about this period taken a direction strongly adverse to emancipation. The Tory reaction which followed the Revolution was still in full force, and a religious movement had been for some time fermenting in England, which had in a great measure dispelled the indifference on doctrinal questions that had long been prevalent, and had greatly intensified the Protestant feeling among the people.

It will be sufficiently evident to anyone who follows the history of the two Churches that their separation had reached its extreme limit when the Puritans were dominant in England and Bossuet was ruling public opinion in France. The Puritans represented Protestantism in its most exaggerated and undiluted form; while Bossuet, who exercised a greater influence over the lay mind than perhaps any theologian since Calvin, was maintaining the tenets of his Church with the most unflagging zeal. He was indeed so far from adopting any extreme or Ultramontane opinions that he even entered into a correspondence with Leibnitz on the possibility of a compromise; but he asserted most emphatically the great distinctive principle of authority; he defined the points of difference with such a rigid accuracy that no evasion was possible; and he laid a greater stress upon dogmas as distinguished from morals than perhaps any other popular writer of his Church. After this period, for about a century, the two systems seemed rapidly approximating. If we compare the sermons of Massillon with those of Bossuet we see the change in its commencement; if we compare the sermons of Blair or of Kirwan with those of the early Anglican divines, we see it in its completion. Dogma had formerly almost superseded practical teaching, but it now in its turn gave way. The Christian preacher became at last simply an expounder of morals. A well-regulated disposition, a virtuous life, and an active benevolence, were represented as almost a summary of Christianity. The Bible was regarded as a repository of noble maxims and of instructive examples. The triumph of religion would be merely the perfection of order, the apotheosis and the completion of government. This tendency may be in part ascribed to the natural reaction and fatigue that followed the fierce controversies of the preceding century; and it was also in a great measure due to the prevalence of scepticism in both Churches. In England sceptical opinions had been maintained openly by Bolingbroke, and Gibbon, and Hume; and if the whole light literature at the close of the last century was not Voltairian in its spirit, it was probably owing in a great measure to the extraordinary influence of Dr. Johnson. In France no such restraint existed. Voltaire and Rousseau towered far above their contemporaries, and never disguised their sentiments. The sarcasms of Voltaire turned the whole stream of ridicule and wit against the Church; while the burning eloquence, the impassioned earnestness, and the intense realising powers of Rousseau, fell with terrific effect on its tottering system. The University of Paris issued an answer to the 'Vicar of Savoy,' but it is now almost forgotten. All the real talent of the country seemed ranged against the established faith, and its defenders were compelled to adopt an apologetical and an evasive tone. It was quite true that all infants who died unbaptised were excluded from heaven, but then hell was an indefinite expression, and comprised a variety of conditions, and St. Augustine was not prepared to say that it would be better for those children had they never been born. Purgatory was undoubtedly a Catholic doctrine, but it was not necessarily the place of torment by fire which was portrayed in the pictures in every church. Though the priests had at one time celebrated almost every royal marriage in Spain by an auto-da-fé, and though a Pope had struck medals in commemoration of the massacre of St. Bartholomew, yet the spirit of Torquemada and of Catherine de' Medicis might be safely reprehended by the orthodox. The doctrine of invincible ignorance was brought prominently forward. The doctrine of infallibility was interpreted in its broadest sense, and the attribute was applied not to an individual, but to the whole Church. Above all, the purity of the moral teaching of Christianity was asserted and displayed, while its special doctrines were allowed to fall into the background. In this manner the two religions began rapidly to assimilate, when the tide again turned, and a violent revulsion set in. In Roman Catholic countries Ultramontanism once more became dominant after the Revolution, but it purchased its triumph dearly. The priests taught the most extreme Roman Catholic doctrines, while the educated laity remained disciples of Montaigne, if not of Voltaire. In England the Methodists had begun their labours; and, after many years of comparatively unnoticed preaching among the poor, their principles began to leaven the higher ranks, and to embody themselves in the great Evangelical party.

The Ultramontane and the Evangelical movements completely altered the attitude of the two religions both towards scepticism and towards each other.

Voltaire had maintained in France that the doctrines of the Church were contrary to reason and to the moral sense; and Ultramontanism answered that these were absolutely incompetent to judge them. Bolingbroke had argued in England that the moral teaching of Christianity existed in the works of the pagan philosophers; and the Evangelical replied that a moral system had no efficacy as a means of salvation, and was only enforced in the New Testament as a secondary and subordinate object. The two sections of Christianity had been approximating, on the ground of common duties; and the Evangelical taught that man could not perform duties acceptably, and that the whole scope and purport of Christianity was to teach a doctrine which the Church of Rome refused to admit. Against this Church, then, as the most powerful, the most subtle, and the most specious opponent of truth, all the energies of the Evangelicals were directed. They traced its lineaments in every intimation of coming apostacy contained in the prophetic writings. They recognised it as the horn of Daniel 'speaking proud things'—as the mystic Babylon, red with the blood of the saints—as the Man of Sin, who was to be revealed when the Roman empire was removed—as the spirit of Antichrist, that was to seduce and to triumph in the latter days. They revived the histories of bygone persecutions that transcended the worst efforts of paganism, and laboured with the same untiring assiduity in the pulpit and on the hustings, in the religious tale and the newspaper article, to repress and to crush the Church they feared.

The Evangelical movement was somewhat slow in spreading to Ireland, and during the greater part of the eighteenth century the Irish Protestant clergy were in general far from bigots. The theological temperature, as I have said, was very moderate, and the habit, which the penal laws produced, of ostensibly passing from one religion to the other in order to join a profession or preserve a property, contributed to lower it. In 1745, it is true, under the fear of an impending invasion, a kind of panic of intolerance passed through the clergy, and they were mischievously active in denouncing the Catholics, but for the most part they were very harmless men, who discharged social and philanthropic functions of unquestionable utility, meddled little with dogmatic theology, and seldom interfered with their Catholic neighbours. The tithe riots of the eighteenth century had little or no connection with religious animosity, and the Protestant landlords were almost as hostile to the tithes as their tenants. In 1725, when the penal laws were at their height, a Protestant clergyman named Synge, in a very remarkable sermon preached before the Irish House of Commons, and published by its order, urged the duty of granting perfect toleration to the Catholics. Ten years later the illustrious Bishop Berkeley, in his 'Querist,' advocated their admission into Dublin University, and their exemption from the obligation of attending chapel or divinity lectures—a policy which was carried out in Ireland near the end of the century. The famous Bishop of Derry was one of the most uncompromising supporters of the Catholic claims. He was, no doubt, too violent and eccentric to be taken as a fair specimen of his order, but the great Relief Bill of 1793, which gave the Catholics the suffrage, was warmly supported by several bishops, and acquiesced in by the majority of the clergy; and it produced nothing of that frantic intolerance which, both among the English and Irish clergy, was aroused by the much less important measure of 1829.[1] Dublin University has always been looked upon as a stronghold of Irish Protestantism, but it was by many years the first university in the kingdom to throw open its degrees to Catholics, and even in the years that followed the Union it was represented by Plunket, at a time when that great orator was leading the Catholic cause.

It would be, I conceive, a mistake to attribute the tolerance of the Irish Protestants towards the close of the eighteenth century to the prevalence of conscious scepticism. Avowed and reasoned free thought has never been very common in Ireland,[2] and the Irish literature towards the end of the last century and the beginning of the present is full of the usual denunciations of scepticism, and the usual depreciation of

---

[1] A contemporary Irish historian thus describes the attitude of the clergy on this occasion: 'What a picture of liberality and moderation did the conduct of the Established clergy of Ireland exhibit during the recent application for Catholic emancipation! Many pious and learned prelates exerted their eloquence in Parliament in support of Catholicity; and the entire body of the Protestant clergy, in their conduct on this occasion, have fully affirmed themselves the disciples of the meek, mild, and gentle Author of Christianity.'—*Mullala's Irish Affairs* (1795), vol. ii. p. 260.

[2] Primate Boulter complained bitterly of 'the growth of atheism, profanity, and immorality' in Ireland, but it seems to have shown itself chiefly in resistance to tithes. Toland was an Irishman, but lived in England, and when he went to Ireland he was denounced from the pulpit, and such an outcry was raised that it became dangerous to speak to him, and he could hardly procure the necessaries of life. He appears however to have been guilty of much imprudence in propagating his views. Parliament ordered his 'Christianity not Mysterious' to be burnt, and the author to be arrested, and he only escaped by precipitate flight. Molyneux has described the transaction in letters to Locke, and South wrote in great glee to the Archbishop of Dublin: 'Your Parliament presently sent him packing, and without the help of a faggot soon made the kingdom too hot for him.'—*Disraeli's Calamities of Authors*, vol. ii. p. 153.

sceptical writers.[1] At the same time the type of prevailing Protestantism, like that of the prevailing Catholicism, was singularly colourless and undogmatic. I have already quoted some sentences from the speeches of Grattan, describing the gradual assimilation of the two creeds, and I may add that no one appears to have been scandalised by the somewhat startling summary of ecclesiastical history which the same speaker threw out in one of his greatest orations: 'The only Divine institution we know of—the Christian religion—did so corrupt as to have become an abomination, and was rescued by Act of Parliament.' In an age when sectarian virulence has obtained a great empire over the minds of men, it seldom fails to reflect itself in the hallucinations of speculators in unfulfilled prophecy; but, as we have already seen, Mr. Dobbs, who was the most enthusiastic Irish labourer in this field, was a warm advocate of the Catholic claims. By far the most eminent man in the Protestant Church at the end of the last century was Dean Kirwan, who, if estimated by the power he exercised over the feelings of his auditors, by the beneficence he evoked, and by the judgments of his contemporaries, at a time when the standard of eloquence was extremely high, must be placed as a pulpit orator almost on a level with Whitefield. This very remarkable man had been originally a Catholic, and one of the reasons he alleged for joining the Established Church was, that he should thus obtain more extensive opportunities of doing good. He rigidly

[1] E.g. 'The writings of Hume and Gibbon, which have been directly or indirectly levelled against the Christian religion, have long since sunk into merited oblivion.'—*Mullala's View of Irish Affairs from the Revolution* (1795), vol. ii. p. 280. 'Surely a Voltaire, a Rousseau, or a Gibbon were as inferior to Colin Maclaurin in mental power as they were in bodily strength to Hercules or Sampson.'—*Ryan's History of the Effects of Religion* (1802), p. 421.

abstained in all his sermons from every topic relating to the differences between the two Churches, making it, as he said, his main object 'to banish religious prejudices, to diffuse through society the great blessings of peace, order, and mutual affection, and to represent Christianity as a practical institution of religion designed to regulate the dispositions and improve the characters of men;' and he at last devoted his talents entirely to pleading the cause of charitable institutions.

A society could not have been very bigoted when it most popular preacher adopted such a tone. Kirwan though a man of spotless reputation and splendid genius, never obtained any more lucrative preferment than a deanery of 400*l.* a year, and was able to leave no fortune to his children; but something of his spirit was shown among his more fortunate brethren. Law, the Bishop of Elphin, was accustomed to distribute among his Catholic parishioners the best books of their own authors, saying that, as he could not make them good Protestants, he at least wished them to be good Roman Catholics.

I have dwelt at some length upon these facts, for they are not much known in England, and they have a considerable importance in the history of public opinion in Ireland. And, indeed, the amount of intolerance that formerly existed in both religions has been not a little exaggerated; for atrocities which were really due to an hostility of races have been ascribed to the conflict of their religions. The Irish have not generally been an intolerant or persecuting people. The early history of the introduction of Christianity into Ireland, though not, as has been said, absolutely bloodless,[1] was at least unusually pacific, and it was an old reproach against Irishmen, that their country, which had pro-

---

[1] There is a discussion on this point in Todd's 'Life of St. Patrick.'

duced innumerable saints, had produced no martyr. During the atrocious persecutions of Mary, the English Protestants were perfectly unmolested in Ireland. The massacre of Protestants in 1642 was so little due to religious causes that the only Englishman of eminence who was treated by the rebels with reverence and care was Bishop Bedell, who was one of the most energetic Protestants of his age, and the first Irish bishop who attempted to proselytise among the Catholics. The Irish people have always been more superstitious than the English, and perhaps than the Scotch, but their superstitions have usually taken a milder form. Many hundreds of unhappy women have perished on the charge of witchcraft both in England and Scotland since the Reformation, but I am not aware of the witch mania having ever raged in Ireland to a degree at all comparable to that in England under James I. and the Puritans, and in Scotland during a great part of the seventeenth century.[1] Whatever animosity the penal laws produced had in a great measure subsided towards the end of the eighteenth century, and it would be difficult to find in any country more moderate or liberal members of their respective faiths than Kirwan, the greatest preacher among the Irish Protestants, and O'Leary, the greatest writer among the Irish Catholics.

The elements of religious animosity, however, though they were almost dormant, existed in abundance, and several causes concurred, with the rise of the Evangelical

---

[1] A famous Irish witch case—that of Dame Alice Kyteler, in 1324—has been reprinted by the Camden Society, and a few unimportant later ones are given by Glanvil in his 'Sadducismus Triumphatus.' Hutchinson, Wright, and Madden appear to have found no other Irish cases. It is much to be wished that some Irish archæological society would investigate more fully than (as far as I am aware) has yet been done the history of Irish witchcraft.

movement, in resuscitating them. The many outbursts of lawless violence that convulsed the country from the middle of the century had been for a long time entirely unconnected with religion. Rack-rents, the fiscal pressure of tithes, the invasions of common land by the landlords, the law which compelled workmen to devote a certain amount of unpaid labour to repairing the county roads, were the causes or pretexts of the appearance of the Whiteboys, the Oakboys, and the Hearts of Steel. In 1785, however, a new type of disturbance began. Protestants in the county Armagh, and afterwards in other districts, began to form bands under the name of Peep-of-Day Boys, and to attack and persecute the Catholics, who then formed societies called 'Defenders,' which were at first a kind of irregular police, and soon after became bands of depredators. The Relief Bill of 1793, conferring votes upon the Catholics, produced some slight economical disturbance; for landlords, who had especially favoured Protestant tenants on account of the political influence they could give, now freely admitted the competition of Catholics. It was not, however, till the furious passions aroused by the recall of Lord Fitzwilliam had broken out that religious animosity became intense. In 1795 the Orangemen came into existence, and signalised themselves by spreading riot over a great part of the north of Ireland. The battle of the Diamond, in which they defeated a large body of Catholics, and in which forty-eight men were killed, took place in the December of this year, and, being sedulously commemorated by the Orangemen, it produced an intense and an enduring animosity. Many Catholics were compelled to emigrate from the county Armagh, and take refuge in Connaught. As the Rebellion became imminent, the violence of sectarian feeling rose to the highest point, and all who tried to allay it were looked upon with suspicion. The name of Grattan was struck off the Privy Council, and the Dublin University authorities removed his picture from their hall, and replaced it by that of Clare. When the Rebellion actually broke out, it aroused all the worst and fiercest passions of the nation. Wesley had before this turned aside from his religious labours to write against the removal of the penal laws. In the Irish Government, Lord Clare was fiercely anti-Catholic, and similar sentiments were energetically maintained in the Irish and afterwards in the English House of Commons by the notorious Dr. Duigenan.

This very singular personage is said to have been himself originally a Roman Catholic. He was a man of low extraction, but of some talents, and had been a Fellow of Trinity College, where he wrote a book against the provost, Hely Hutchinson. He obtained a seat in the Irish House of Commons, and laboured without success to procure the cessation of the Maynooth grant which had been made during the administration of Lord Fitzwilliam. He was one of the warmest supporters of the Union, and in the English Parliament the most vituperative and indefatigable opponent of the Catholic claims. He adopted that method which is still employed by some politicians, of exhuming all the immoral sentiments of the schoolmen, the Jesuit casuists, and the mediæval councils, and parading them continually before the Parliament and before the country.[1] Against this system Grattan energetically protested. 'No religion,' he said in one

[1] It is curious that he was married to a Roman Catholic; he proposed to her and was refused when young, but was accepted many years after, when she was a widow. In spite, however, or perhaps in consequence of matrimony, his antipathy to the Church of Rome continued unabated to the end.

of his speeches, 'can stand if men, without regard to their God, and with regard only to controversy, shall rake out of the rubbish of antiquity the obsolete and quaint follies of the sectarians, and affront the majesty of the Almighty with the impudent catalogue of their devices; and it is a strong argument against the proscriptive system that it helps to continue this shocking contest; theologian against theologian, polemic against polemic, until the two madmen defame their common parent, and expose their common religion.'

Every year the state of feeling in Ireland became worse. As is always the case, the destruction of national feeling gave an increased bitterness to sectarian controversy, and turned almost all the energies of the country into that channel. The Roman Catholics, who had formerly been almost passive, began to agitate vehemently, and to complain bitterly that Pitt opposed their emancipation, though he had formerly professed himself favourable to it. The Evangelical movement in Ireland had chiefly assumed an aggressive character, and the effects of the Rebellion of '98 had not yet subsided. A few years after the Union there were no less than five distinct parties agitating actively: the French party, who cherished the traditions of '98; the armed Orangemen, who were pillaging in the county Armagh; the more pacific Tories, who were arguing against emancipation; the moderate Liberals, who followed Grattan, and comprised a large section of the Protestants, and almost all the higher orders of Roman Catholics; and the clerical and democratic party, which was beginning to rise under the inspiration of O'Connell. When we add to this that the English public was becoming thoroughly permeated by the Evangelical movement, the difficulty of Grattan's position becomes very apparent.

He determined to keep himself entirely independent. He refused office in Fox's Ministry, which came in in 1806, and he refused to accept 4000*l*. which the Roman Catholics subscribed in the same year to defray the expenses of his election for Dublin. He kept up a correspondence with every section of the Constitutional Liberals, but he would not place himself in the hands of any. In 1807 he incurred much unpopularity by supporting the Government Coercion Bill, which he believed to be necessary on account of the disorganised condition of the country.[1] In 1808 he entered into the Veto question. This proposition, which at one time created so much agitation, was an attempt to produce a compromise; the English Parliament consenting to emancipate the Catholics, on the condition that a power of veto was reserved to the English Sovereign in the election of Catholic bishops. The proposal was then much discussed and warmly accepted by the whole body of Roman Catholics of England, by the upper order of those of Ireland, and by Grattan himself. The Court of Rome was very conciliatory, and the Irish bishops in 1808, by the agency of Dr. Milner, declared their willingness to accept it; but they soon yielded to the popular outcry and to the influence of O'Connell, and under the leadership of the same prelate vehemently opposed it. This produced a complete schism between the gentry and the clergy, and undoubtedly retarded the triumph of the cause. In 1813 a Bill, accompanied by the veto and some minor securities, actually passed a second reading, and was finally rejected by a majority of only four, but the bishops afterwards

---

[1] He said he hoped to secure to Ireland a 'reversionary interest in the constitution.' He adopted a similar course in 1814. The perfect courage with which Grattan always risked his popularity for what he thought the interest of his country is one of the finest traits of his character.

denounced it. In the following year the Catholic Board, at the suggestion of O'Connell, called upon Grattan to place himself under their direction, and upon his refusal took their petition out of his hands, and entrusted it to Sir Henry Parnell.

It was touching to see the old statesman thus superseded in the cause he had served so long, yet rising without one word of complaint, of recrimination, or of bitterness, to support his younger colleague. The more moderate party still made him their representative, and nothing in his whole career is more admirable than the good taste and the self-abnegation which he manifested throughout. He made it a rule, as he said, 'never to defend himself at the expense of his country,' and he displayed the same zeal and the same eloquence as when his popularity was greatest. The ill-feeling was at one time so strong that, after his election for Dublin in 1818, he was assaulted by a mob in the streets. All parties were heartily ashamed of the act, and the Roman Catholics and the Orangemen reciprocally charged each other with the guilt.[1] Notwithstanding this ebullition, there can be little doubt that he rose higher and higher in the estimation of the educated of all parties, and that the moderation and the exquisite tact he manifested exercised a most powerful influence upon Parliament. O'Connell adopted an entirely different course; but, as we shall see, O'Connell's object was, in all probability, a different one; and even when opposing Grattan, he extolled his patriotism in the highest terms. A living historian has noticed, on the authority of Sir R. Peel, a curious indication of the veneration with which

---

[1] Grattan himself, when asked by some English friends about the cause of the riot, answered: 'It was religion—it was religion—and religion broke my head.'

Grattan was at this time regarded The members who had sat with him in the Irish House of Commons were accustomed in the English House always to address him with a 'Sir,' as they would the Speaker, and this custom was followed by Lord Castlereagh at a time when he was the leader of the House.[1]

To the Catholic question Grattan devoted the entire energies of his latter years. With the exception of one very brilliant and very successful speech in favour of immediate war with France, in 1815, he never spoke at length on any other subject. In 1819 he was defeated by a majority of only two; and in 1820 he went over to London, to bring the subject forward again, when the illness under which he had for some time been labouring assumed a more violent and deadly character. He lingered for a few days, retaining to the last his full consciousness and interest in public affairs. Those who gathered around his death-bed observed with emotion how fondly and how constantly his mind reverted to that Legislature which he had served so faithfully and had loved so well. It seemed as though the forms of its guiding spirits rose more vividly on his mind as the hour approached when he was to join them in another world; and, among the last words he is recorded to have uttered, we find a warm and touching eulogium of his great rival, Flood, and many glowing recollections of his fellow-labourers in Ireland. He passed away tranquilly and happily on June 6, 1820. He died, as a patriot might wish to die, crowned with honours and with years, with the love of friends and the admiration of opponents, leaving a nation to deplore his loss and not an enemy to obscure his fame.

It is at the tombs of great men that succeeding

[1] Lord Mahon's 'History of England.'

generations kindle the lamp of patriotism; and it might have been supposed that he whose life was fraught with so many weighty lessons, and whose memory possesses so deep a charm, would have rested at last in his own land and among his own people. Another, and, as it would seem to some, a nobler lot, was reserved for Grattan. A request was made to his friends that his remains might rest in Westminster Abbey, and that request was complied with. He lies near the tombs of Pitt and Fox. The place is an honourable one, but it was the only honour that was bestowed on him. Not a bust, not an epitaph marks the spot where the greatest of Irish orators sleeps; but one stately form seems to bend in triumph over that unnoticed grave. It is the statue of Castlereagh, 'the statesman of the legislative Union.'

DANIEL O'CONNELL.

WHILE the Union was under discussion in the Irish Parliament no class of persons exerted themselves more energetically in opposing it than the Dublin lawyers. Among the meetings they held for this purpose there was one which assumed a peculiar significance from its being composed entirely of Roman Catholics. They assembled to protest against the assertion that the Roman Catholics, as a body, were favourable to the measure; to express their opinion that it would exercise an injurious influence upon the struggle for emancipation; and to declare that were it otherwise they did not desire to purchase that boon at the expense of the independence of the nation. Military law was then reigning, and a body of troops, under Major Sirr, were present at the Exchange to watch the proceedings. It was under these rather trying circumstances that a young lawyer, 'trembling,' as he afterwards said, 'at the sound of his own voice,' rose to make his maiden speech. He delivered a short address against the Union, which, if it contained no very original or striking views, had at least the merit of exhibiting the common arguments in the clearest and most convincing light; and he shortly after hurried to a newspaper-office to deposit a copy for publication. This young lawyer was Daniel O'Connell, the great Irish agitator. I confess that it is not without some hesitation that I approach this part of my subject, for the difficulty of painting the character of O'Connell with fairness and impartiality can hardly be exaggerated. 'Never, perhaps,' as has been said, 'was there a man at once so hated and so loved;' and it may be doubted whether any public man of his time was the object of so much extravagant praise and blame. On the whole, however, the latter greatly preponderates. For many years the entire press of England, and a large section of that of Ireland, was ceaselessly employed in denouncing him. All parties in England were combined against him, and in Parliament he had to bear alone the assaults of statesmen and of orators of the most varied opinions. By the more violent Irish Protestants he was regarded with feelings of mingled hatred and terror that almost amounted to a superstition; and the failure of the last great struggle of his life, as well as the disastrous condition of the country at the time of his death, has been very injurious to his reputation.

Daniel O'Connell was born in the county of Kerry, in the year 1775. His family was one which had for a long time occupied a prominent position among the Catholics of the county, which was much noted for its national feeling, and, it must be added, greatly addicted to smuggling. It was in after-years remarked as a curious coincidence that its crest bore the proud motto 'Oculus O'Connell Salus Hiberniæ.' During his boyhood the penal laws were still unrepealed, though much relaxed in their stringency, and the poorer Roman Catholics had sunk into that state of degradation which compulsory ignorance necessarily produces, while the richer drew their opinions, with their education, from France. O'Connell spent a year at St. Omer, where the principal predicted that he would afterwards distinguish himself, and he then remained for a few months at the English College of Douay. The Revolution had at this time shattered the French Church and crown, and the minds of all men were violently agitated in its favour or against it. O'Connell's sympathies were strongly opposed to the movement. Like the members of most Irish families that had adhered to their religion during the penal laws, he was deeply attached to it, politically and through feelings of honour, if not from higher motives. Besides this, the associations of his college were necessarily clerical; and some of the revolutionary soldiers, in passing through Douay, had heaped many insults on the students. On his return to Ireland he found that the contagion of the Revolution had already spread, and in the year '98, when he was called to the Bar, rebellion was raging over the country. He became a member of a yeomanry corps which the lawyers had formed, and was at that time, as he afterwards confessed, 'almost a Tory.' Though he retained to the last his antipathy to rebellion, his opinions in other respects were soon altered by the scandalous scenes of the State trials, by the spectacle of the condition of his co-religionists, and above all by the circumstances attending the Union.

The Roman Catholics had made some inconsiderable efforts to influence public opinion by a society for the purpose of preparing petitions for Parliament, and of this society he early became a member. His extraordinary eloquence, his fertility of resources, his sagacity in reading characters and in discerning opportunities, his boundless and ever daring ambition, soon made him the life of this society, and outweighed all the advantages of rank and old services that were sometimes opposed to his views. There is much reason to believe that almost from the commencement of his career he formed one vast scheme of policy which he pursued through life with little deviation, and, it must be added, with little scruple. This scheme was to create and lead a public spirit among the Roman Catholics; to wrest emancipation by this means from the Government; to perpetuate the agitation created for that purpose till the Irish Parliament had been restored; to disendow the Established Church; and thus to open in Ireland a new era, with a separate and independent Parliament and perfect religious equality. It would be difficult to conceive a scheme of policy exhibiting more daring than this. The Roman Catholics had hitherto shown themselves absolutely incompetent to take any decisive part in politics. They were not, it is true, quite as prostrate as they had been when Swift so contemptuously described them as being 'altogether as inconsiderable as the women and children, ... without leaders, without discipline, without natural courage, little better than hewers of wood and drawers of water, and out of all capacity of doing any mischief if they were ever so well inclined;' but yet the iron of the penal laws had entered into their souls, and they had always thrown themselves helplessly on Protestant leaders. Grattan, it is true, was now in the decline of life, but Plunket, who was still in the zenith of his great powers, was ready to succeed him. If the Roman Catholics could be braced up to independent exertion the noblemen and men of property in their ranks would be their natural leaders, and, at all events, a young lawyer, dependent on his talents and excluded from Parliament and from the higher ranks of his profession, would seem utterly unfitted for such a position. O'Connell, however, perceived that it was possible to bring the whole mass of the people into the struggle, and to give them an almost unexampled momentum and unanimity by applying to politics a great power that lay dormant in Ireland—the power of the Catholic priesthood. To make the priests the rulers of the country, and himself the ruler of the priests, was his first great object.

Few things are more striking to those who compare the present condition of Ireland with her past than the rapidity with which the power of the priests has augmented during the present century. Formerly they were much loved by their flocks but much despised by the Protestants, and they were contented with keeping alive the spiritual feeling of their people without taking any conspicuous part in politics. Once or twice, indeed, the bishops came forward to disclaim certain doctrines that were attributed to their Church, and were advanced as an argument against emancipation. Once or twice they held meetings to further the movement by expressing their willingness to concede something to procure the boon. If they had taken a certain part in favour of the Union, it was at the desire of the Ministers, and the position of O'Leary was solely due to the extreme beauty of his style.

Strange as it may now appear, the priests seem to have been at one time most reluctant to enter into the political arena, and the whole agitation was frequently in danger of perishing from very languor. There was a party supported by Keogh, the leader in '93, who recommended what was called 'a dignified silence'—in other words, a complete abstinence from petitioning and agitation. With this party O'Connell successfully grappled. His advice on every occasion was, 'Agitate, agitate, agitate!' and Keogh was so irritated by the defeat that he retired from the society. But the greatest of the early triumphs of O'Connell was on the Veto question. It is evident that if the proposed compromise were made, the policy he had laid out for himself would be completely frustrated. A public spirit would not be formed among the Roman Catholics by a protracted struggle. Emancipation would be a boon that was conceded, not a triumph that was won; and the episcopacy would be in a measure dependent upon the Crown. In the course of the contest almost every element of power seemed against him. The bishops, both in 1799 and in 1808, had declared themselves in favour of the veto. The English Roman Catholics led by Mr. Butler, the upper order of those of Ireland led by Lord Fingall, and the Protestant Liberals led by Grattan, warmly supported it. Sheil, who was thoroughly identified with the democratic party, and whose wonderful rhetorical powers gave him an extraordinary influence, wrote and spoke in favour of compromise; and, to crown all, Monsignor Quarantotti, who in a great measure managed affairs at Rome during the captivity of Pius VII., exhorted the bishops to accept it. Over all these obstacles O'Connell triumphed. He succeeded in persuading or forcing the bishops into violent opposition to the scheme, and in throwing them on the support of the people. Dr. Milner wrote against the veto, and was accordingly censured by the English Roman Catholics; but O'Connell induced those of Ireland to support him. Grattan refused to place himself in the hands of the Catholic committee, and the petition was immediately taken out of his hands. Lord Fingall, Sir E. Bellew, and a few other leading Catholics, would not yield, and were obliged to form a separate society, which soon sank into insignificance. Sheil was answered by O'Connell, and the answer was accepted by the people as conclusive; and, finally, the rescript of Quarantotti was disobeyed by the bishops and disavowed by the Pope. The results of the controversy were probably by no means beneficial to the country, but they at least served in an eminent degree the purposes of the agitator. The clergy were brought actively into politics. The lower orders were stirred to the very depths, and O'Connell was triumphant over all rivals.

In the course of this controversy it was frequently urged that O'Connell's policy retarded emancipation. This objection he met with characteristic frankness. He avowed himself repeatedly to be an agitator with an 'ulterior object,' and declared that that object was the repeal of the Union. 'Desiring, as I do, the repeal of the Union,' he said in one of his speeches, in 1813, 'I rejoice to see how our enemies promote that great object. Yes, they promote its inevitable success by their very hostility to Ireland. They delay the liberties of the Catholics, but they compensate us most amply because they advance the restoration of Ireland. By leaving one cause of agitation, they have created, and they will embody and give shape and form to, a public mind and a public spirit.' In 1811, at a political dinner, he spoke to the toast of Repeal, which had been given at his suggestion, and he repeatedly reverted to the subject. Nothing can be more untrue than to represent the Repeal agitation as a mere afterthought designed to sustain his flagging popularity. Nor can it be said that the project was first started by him. The deep indignation that the Union had produced in Ireland was fermenting among all classes, and assuming the form, sometimes of a French party, sometimes of a social war, and sometimes of a constitutional agitation. The Repeal agitation directed, but did not create, the national feeling. It merely gave it a distinct form, a steady action, and a constitutional character. In 1810 a very remarkable movement in this direction took place in Dublin. The grand jury passed a resolution declaring that 'the Union had produced an accumulation of distress; and that, instead of cementing, they feared that if not repealed it might endanger the connection between the sister countries.' In the same year a meeting communicated on the subject with Grattan, who was member for the city. Grattan replied that a Repeal agitation could only be successful if supported by the people; that if that support were given, he would be ready to advocate the movement; and that he considered such a course perfectly consonant with devoted attachment to the connection.[1] Lord Cloncurry relates that he was a member of a deputation which on another occasion waited on Grattan, and that Grattan said to them, 'Gentlemen, the best advice I can give my fellow-citizens upon every occasion is to keep knocking at the Union.'

The prominent position O'Connell had assumed in politics naturally exercised a favourable influence upon his professional career, so that he became by far the most popular counsel in Ireland, and was invariably employed in all those cases which involved political

---

[1] Grattan's letter is so remarkable that I give it in full. It will be found in his Life, by his son:

'Gentlemen,—I had the honour to receive an address, presented by your committee, and expressive of their wishes that I should present certain petitions and support the repeal of an Act entitled the "Act of Union," and your committee adds, that it speaks with the authority of my constituents, the freemen and freeholders of the City of Dublin. I beg to assure your committee, and through them my much beloved and much respected constituents, that I shall accede to their proposition. I shall present their petitions and support the repeal of the Act of Union with a decided attachment to our connection with Great Britain, and to that harmony between the two countries, without which the connection cannot last. I do not impair either, as I apprehend, when I assure you that I shall support the repeal of the Act of Union. You will please to observe that a proposition of that sort in Parliament, to be either prudent or possible, must wait until it should be called for and backed by the nation. When proposed, I shall then, as at all other times I hope I shall, prove myself an Irishman, and that Irishman whose first and last passion was his native country.'

or religious considerations. There have been a few lawyers of deeper knowledge, and even of more powerful eloquence, though he ranked extremely high in both respects; but never, perhaps, was there a man more admirably calculated to excel at the Irish Bar. His unrivalled knowledge of the Irish character; his sagacity in detecting the weaknesses of the judges, jurymen, and witnesses; the wonderful dexterity with which he could avail himself of any legal quibble or ambiguity; and the unblushing audacity with which he could confront any opponent, enabled him quickly to distance all competitors. It is difficult for those who are habituated only to the law-courts of England to conceive the vast difference in this respect between the two countries. The difference of the characters of the two nations is nowhere more apparent, and, besides this, the proceedings of the Irish law-courts have ever been deeply tinged with religious and political considerations. In appointments of judges and of law-officers the first question asked by the public seems to be their religion, the second their politics, the last their legal knowledge; and the scandal of mere party judges has been both more frequent and more recent in Ireland than in England. Besides this, an unusual proportion of the leading politicians of Ireland have been practising barristers, and the temptation of making a trial on a question of tithes, or tenant-right, or libel, an occasion for a brilliant display, was irresistible both to the politician and to the orator. As trials of this nature were continually occurring, and as their exclusion from the inner bar and from the bench gave the Roman Catholics a tenfold virulence, the scenes which took place at the Four Courts during the earlier part of the century may be more easily conceived than described. O'Connell always defended the excessive violence of his language, both at the Bar and on the platform, on the ground of the peculiar position of the Roman Catholics. He said that he had found his co-religionists as broken in spirit as they were in fortune; that they had adopted the tone of the weakest mendicants; that they seemed ever fearful of wearying the dominant caste by their importunity, and that they were utterly unmindful of their power and of their rights. His most difficult task was to persuade them of their strength, and to teach them to regard themselves as the equals of their fellow-countrymen. The easiest way of breaking the spell was to adopt a defiant and an overbearing tone. The spectacle of a Roman Catholic fearlessly assailing the highest in the land with the fiercest invective and the most unceremonious ridicule, was eminently calculated to invigorate a cowering people. A tone of extreme violence was the best corrective for a spirit of extreme servility.

There is undoubtedly some truth in these considerations, and they extenuate not a little the language of O'Connell; but they are certainly far from justifying it, either morally or politically. The ceaseless torrent of the coarsest abuse which at every period of his life, and in every sphere in which he moved, he poured upon all opponents; the rapidity with which he passed, on a very small provocation, from a tone of the most hyperbolical praise to language that was worthy of Billingsgate; and the virulence with which he attacked some of the most illustrious characters in the country, prejudiced all moderate men against him. It was said of him that his mind consisted of two compartments—the one inhabited by the purest angels, and the other by the vilest demons—and that the occupation of his life was to transfer his friends from the one to the other. A man who did not hesitate to describe the Duke of Wellington as 'a stunted corporal,' and who applied to other opponents such terms as 'a mighty big liar,' or 'a lineal descendant of the impenitent thief,' or 'a titled buffoon,' or 'a contumelious cur,' or 'a pig,' or 'a scorpion,' or 'an indescribable wretch,' placed himself beyond the pale of courtesy. The abuse he at one period of his life poured upon the Whigs embarrassed him during all the later part of his career, and he drew down upon himself the formal reprimand of the House of Commons by accusing the Tory members on election committees of 'foul perjury.' Such language could hardly fail to lower the character of the movement, and it especially weakened his position when he became a member of Parliament. That tone of gentlemanly moderation, that well-bred, pungent raillery which is so characteristic of the English Parliament, and has been brought to the greatest perfection by Lord Palmerston, has often proved a more efficient weapon of debate than the most splendid eloquence or the most trenchant wit. It draws a magic circle around the speaker, which only similar weapons can penetrate, and it seldom fails to secure the attention and the respect of the public.

The greatest speeches of O'Connell at the Bar were in defence of Magee, the editor of the 'Evening Post,' who had libelled the Duke of Richmond. They consist chiefly of an invective against Saurin, the Attorney-General, as the representative of the Orange party, and were so violent that the publication of one of them was pronounced to be an aggravation of the original libel. In point of eloquence, however, they rank very high; but they are almost exclusively political, for the case of his client was a hopeless one. The principal success of O'Connell at the Bar was, perhaps, not in oratory, but in cross-examining. He had paid special attention to this department, which naturally fell, in a great measure, to the Roman Catholic lawyers at a time when they were excluded from the inner bar; and he brought it to a degree of perfection almost unparalleled in Ireland. His wonderful insight into character, and tact in managing different temperaments, enabled him to unravel the intricacies of deceit with a rapidity and a certainty that seemed miraculous, and his biographies are full of almost incredible illustrations of his skill![1]

It would be tedious to follow into minute detail the difficulties and the mistakes that obstructed the Catholic movement, and were finally overcome by the energy or the tact of O'Connell. For some time the gravest fears were entertained that the Pope would pronounce in favour of the veto. A strong party at Rome, headed by Cardinal Gonsalvi, was known to advocate it, and the deputy of the Irish bishops adopted so importunate a tone that he was peremptorily dismissed, and pronounced by his Holiness to be 'intolerable.' Innumerable dissensions dislocated the movement, and demanded all the efforts of O'Connell to appease them. When the Roman Catholic gentry had seceded, a multitude of those eccentric characters who are ever ready to embark in agitation from the mere spirit of adventure assumed a dangerous prominence, and it was found necessary to adopt a most despotic tone to repress them. The hopes that were entertained of the Prince of Wales produced a great deal of gross and vulgar flattery, and in 1812, when the change in his sentiments became known, some most injudicious resolutions, ascribing it to 'the fatal witchery of an unworthy

---

[1] See especially Mr. O'Neil Daunt's very interesting 'Personal Recollections of O'Connell.'

secret influence.' When he visited Ireland after his coronation, the unbounded sycophancy of some of the Orangemen on one side, and of O'Connell on the other, went far to justify the somewhat strange saying of Swift, that 'loyalty is the foible of the Irish.' Lord Byron, who took a strong interest in the Catholic cause, which he defended in the House of Lords, was justly indignant, and branded the conduct of O'Connell with great severity in the 'Irish Avatar.'

In 1815 O'Connell fought a duel with a gentleman named D'Esterre, which was attended by some very painful circumstances, and gave rise to much subsequent discussion. It arose out of the epithet 'beggarly' which O'Connell had applied to the corporation of Dublin. D'Esterre was killed at the first shot. In the same year Mr. Peel had challenged O'Connell, on account of some violent expressions he had employed. O'Connell, however, was very opportunely arrested at his wife's information, and bound over to keep the peace.

Several times the movement was menaced by Government proclamations and prosecutions. Its great difficulty was to bring the public opinion of the whole body of the Roman Catholics actively and habitually into the question. The skill and activity of O'Connell in arousing the people were beyond all praise, and the consciousness of the presence of a great leader began to spread through the whole mass of the ignorant, dispirited, and dependent Catholics. All preceding movements since the Revolution (except the passing excitement about Wood's halfpence) had been chiefly among the Protestants or among the higher order of the Catholics. The mass of the people had taken no real interest in politics, had felt no real pain at their disabilities, and were politically the willing slaves of their landlords. For the first time, under the influence of O'Connell, the great swell of a really democratic movement was felt. The simplest way of concentrating the new enthusiasm would have been by a system of delegates, but this had been rendered illegal by the Convention Act. On the other hand, the right of petitioning was one of the fundamental privileges of the constitution. By availing himself of this right O'Connell contrived, with the dexterity of a practised lawyer, to violate continually the spirit of the Convention Act, while keeping within the letter of the law. Proclamation after proclamation was launched against his society, but by continually changing its name and its form he generally succeeded in evading the prosecutions of the Government.

These early societies, however, all sink into insignificance compared with that great Catholic Association which was formed in 1824. The avowed objects of this society were to promote religious education, to ascertain the numerical strength of the different religions, and to answer the charges against the Roman Catholics embodied in the hostile petitions. It also *recommended* petitions (unconnected with the society) from every parish, and aggregate meetings in every county. The real object was to form a gigantic system of organisation, ramifying over the entire country, and directed in every parish by the priests, for the purpose of petitioning and in every other way agitating in favour of emancipation. The Catholic Rent was instituted at this time, and it formed at once a powerful instrument of cohesion and a faithful barometer of the popular feeling. It is curious that at the first two meetings O'Connell was unable to obtain the attendance of ten members to form a quorum. On the third day the same difficulty at first occurred, but O'Connell at length induced two Maynooth students who were passing to make up the requisite number, and the introduction of this clerical element set the machine in motion. Very soon, however, the importance of the new society became manifest. Almost the whole priesthood of Ireland were actively engaged in its service, and it threatened to overawe every other authority in the land. In the elections of 1826 sacerdotal influence was profoundly felt; and the defeat of the Beresfords in the Catholic county of Waterford, in which, in spite of their violent anti-Catholicism, they had for generations been supreme, foreshadowed clearly the coming change. The people were organised with unprecedented rapidity, and O'Connell and Sheil traversed the country in all directions to address them.

Though both were marvellously successful in swaying and in fascinating the multitude, it would be difficult to conceive a greater contrast than was presented by their styles.

Richard Lalor Sheil forms one of the many examples of splendid oratorical powers clogged by insuperable natural defects. His person was diminutive, and wholly devoid of dignity; his voice shrill, harsh, and often rising into a positive shriek; his action, though indicative of an intense earnestness, violent without gracefulness, and eccentric even to absurdity. He had distinguished himself as a poet and a dramatist; and it was, perhaps, in consequence of the habits he acquired in those fields that his speeches, though extremely beautiful as compositions, were always a little overcharged with ornament, and a little too carefully elaborated. They seem exactly to fulfil Burke's description of perfect oratory, 'half poetry, half prose;' yet we feel that their ornaments, however beautiful in themselves, offend by their profusion. Two very high excellences he possessed to a pre-eminent degree—the power of combining extreme preparation with the greatest passion, and of blending argument with declamation. There are very few speakers from whom it would be possible to cite so many passages with all the sustained rhythm and flow of declamation, yet consisting wholly of condensed arguments. He was a great master of irony, and, unlike O'Connell, could adapt it either to a vulgar or to a refined audience. He had but little readiness, and almost always prepared the language as well as the substance of his speeches; but he seems to have carefully followed the example of Cicero in studying the case of his opponents as fully as his own, and was thus enabled to anticipate with great accuracy the course of the debate. He was more calculated to please than to move, and to dazzle than to convince.

In almost every respect O'Connell differed from Sheil. Had he been a man of second-rate talent, he would have imitated some of the great orators who adorned the Irish Parliament; he would have studied epigram like Grattan, or irony like Plunket, or polished declamation like Curran. He seems, however, to have early felt that neither the character of his mind nor the career he had chosen were propitious for these forms of eloquence, while he was eminently fitted to excel in other ways. He possessed a voice of almost unexampled perfection. Rising with an easy and melodious swell, it filled the largest building and triumphed over the wildest tumult, while at the same time it conveyed every inflection of feeling with the most delicate flexibility. It was equally suited for impassioned appeal, for graphic narration, and for sweeping the finer chords of pathos and of sensibility. He had studied carefully that consummate master of elocution William Pitt.

and he had acquired an almost equal skill. No one knew better how to pass from impetuous denunciation to a tone of subdued but thrilling tenderness. No one quoted poetry with greater feeling and effect; no one had more completely mastered the art of adapting his voice to his audience, and of terminating a long sentence without effort and without feebleness. His action was so easy, natural, and suited to his subject, that it almost escaped the notice of the observer. His language was clear, nervous, and fluent, but often incorrect, and scarcely ever polished. Having but little of the pride of a rhetorician, he subordinated strictly all other considerations to the end he was seeking to achieve, and readily sacrificed every grace of style in order to produce an immediate effect. 'A great speech,' he used to say, is a very fine thing; but, after all, the verdict is THE thing.' As Sheil complained, 'he often threw out a brood of sturdy young ideas upon the world without a rag to cover them.' He had no dread of vulgar expressions, coarse humour, or undignified illustrations; but at the same time he seldom failed to make a visible impression; for, in addition to the intrinsic power of his eloquence, he possessed in the highest degree the tact which detects the weaknesses and prejudices of his audience and the skill which adapts itself to their moods. His readiness in reply was boundless, his arguments were stated with masterly force, and his narrative was always lucid and vivid. If he endeavoured to become eloquent by preparation, he grew turgid and bombastic; if he relied exclusively on the feelings of the moment, he often rose to a strain of masculine beauty that was all the more forcible from its being evidently unprepared. His bursts of passion displayed that freshness and genuine character that art can so seldom counterfeit.

The listener seemed almost to follow the workings of his mind—to perceive him hewing his thoughts into rhetoric with a negligent but colossal grandeur; with the chisel, not of a Canova, but of a Michael Angelo. Were we to analyse the pleasure we derive from the speeches of a brilliant orator, we should probably find that one great source is this constant perception of an ever-recurring difficulty skilfully overcome. With some speakers appropriate language flows forth in such a rapid and unbroken stream that the charm of art is lost by its very perfection. With others the difficulties of expression are so painfully exhibited or so imperfectly overcome that we listen with feelings of apprehension and of pity. But when the happy medium is attained—when the idea that is to be conveyed is present for a moment to the listener's thought before it is moulded into the stately period—the music of each balanced sentence acquires an additional charm from our perception of the labour that produced it. In addressing the populace the great talents of O'Connell shone forth with their full resplendency. Such an audience alone is susceptible of the intense feelings the orator seeks to convey, and over such an audience O'Connell exercised an unbounded influence. Tens of thousands hung entranced upon his accents, melted into tears or convulsed with laughter—fired with the most impassioned and indignant enthusiasm, yet so restrained that not an act of riot or of lawlessness, not a scene of drunkenness or of disorder, resulted from those vast assemblies. His genius was more wonderful in controlling than in exciting, and there was no chord of feeling that he could not strike with power. Other orators studied rhetoric—O'Connell studied man.

If we compare the two speakers, I should say that before an uneducated audience O'Connell was wholly unrivalled, while before an educated audience Sheil was most fitted to please and O'Connell to convince. Both were powerful reasoners, but the arguments of O'Connell stood in bold and clear relief, while the attention was somewhat diverted from those of Sheil by the ornaments and mannerism that accompanied them. Both possessed great powers of ridicule, but in O'Connell it assumed the form of coarse but genuine humour, and in Sheil of refined and pungent wit. By too great preparation Sheil's speeches displayed sometimes an excess of brilliancy. By elaborate preparation O'Connell occasionally fell into bombast. O'Connell was much the greater debater, Sheil was much the greater master of composition. O'Connell possessed the more vigorous intellect, and Sheil the more correct taste.

The success of the Catholic Association became every week more striking. The rent rose with an extraordinary rapidity. The meetings in every county grew more and more enthusiastic, the triumph of priestly influence more and more certain. The Government made a feeble and abortive effort to arrest the storm by threatening both O'Connell and Sheil with prosecution for certain passages in their speeches. The sentence cited from O'Connell was one in which he expressed a hope that 'if Ireland were driven mad by persecution a new Bolivar might arise,' but the employment of this language was not clearly established, and the Bill was thrown out. The speech which was to have drawn a prosecution upon Sheil was a kind of dissertation upon 'Wolfe Tone's Memoirs,' of which Canning afterwards said that it might have been delivered in Parliament without even eliciting a call to order. The Attorney-General was Plunket, who by this act completed the destruction of his influence in Ireland. Sheil asked him, as a single favour, to conduct the prosecution in person. Had he done so, Sheil intended to cite the passages from Plunket's speeches on the Union, which at least equalled in violence any that the Repealers ever delivered. The dissolution of the Government prevented the intended prosecution.

One very serious consequence of the resistance to the demand for emancipation was the strengthening of the sympathy between Ireland and France. The French education of many of the Irish priests, and the prominent position of France among Roman Catholic nations, had naturally elicited and sustained it. The sagacity of O'Connell readily perceived what a powerful auxiliary foreign opinion would be to his cause; and by sending the resolutions of the association to Catholic Governments, by translations of the debates, and by a series of French letters written by Sheil, the feeling was constantly fanned. Many Irishmen have believed that the existence of this sympathy is an evil. I confess I can hardly think so. Irishmen should never forget how, in the hour of their deepest distress, when their energies were paralysed by a persecuting code, and their land was wasted by confiscation and war, France opened her ranks to receive them, and afforded them the opportunities of honour and distinction they were denied at home. Gratitude to the French nation is a sentiment in which both Irish Catholics and Irish Protestants may cordially concur. The first will ever look back with pride to the achievements of the Irish brigade, which threw a ray of light over the gloomiest period of their depression. The second should not wholly forget that to the enterprise of French refugees is due a large part of the manufactures which constitute a main element of their prosperity. Nor is it possible for any patriotic Irishman to contrast without emotion the tone which has been adopted towards his country by some of the most eminent writers of France with the studied depreciation of the Irish character by some of the most popular authors and by a large section of the press of England. The character of a nation is its most precious possession, and it is to such writers as Montalembert and Gustave de Beaumont that it is mainly due that Ireland has still many sympathisers on the Continent.

But in addition to these considerations there are others of much weight that may be alleged. One of the most important intellectual advantages of Catholicism is, that the constant international communication it produces corrects insular modes of thought, and it has been of no small benefit to Irishmen that they have never been altogether without some tincture of French culture. In the worst period of the last century this was secured by the French education of the priests; and, in spite of geographical position and of penal laws, a certain current of continental ideas has always been perceptible among the people. The spirit of French Catholicism long gave a larger and more liberal character to Irish Catholicism, and in French literature Irish writers have found the supreme models of a type of excellence which is peculiarly congruous to the national mind. There have sometimes been political dangers arising from the sympathy between the nations; but on the whole it has, I believe, produced far more good than evil.

The formation of the Wellington Ministry seemed effectually to crush the present hopes of the Catholics, for the stubborn resolution of its leader was as well known as his Tory opinions. Yet this Ministry was destined to terminate the contest by establishing the principle of religious equality. The first great concession was won by Lord J. Russell, who, by obtaining the repeal of the Corporation and Test Acts, secured the admission of Dissenters to the full privileges of the constitution. The Tory theory that the State having an established religion, the members of that religion had a right to a position of political ascendency, was thus for the first time rejected, and with it fell the most popular argument against Catholic emancipation. O'Connell and the Catholics warmly supported the Dissenters in their struggle for emancipation, but the 'No Popery' feeling among the latter was so strong that they never reciprocated the assistance. Even at a time when they were themselves suffering from disabling laws, they were in general hostile to Catholic emancipation.

About this time a new project of compromise was much discussed, both in Parliament and by the public, which shows clearly how greatly the prospects of the cause had improved. This project was, that the emancipation should be accompanied by the payment of the clergy by the State, and by the disfranchisement of the 40s. freeholders. It seems to have been very generally felt that while emancipation could not be long delayed, some measure should be taken to prevent the Roman Catholic body from being virtually independent of the Crown. It was felt that a body which was connected by interests, by sympathies and allegiance, with a foreign Court, might become very dangerous in Parliament. To pay the Roman Catholic clergy would be to unite them by a strong tie to England, and to place them in a measure under the control of the Government. It would also, in all probability, set at rest the long-vexed question of the Established Church. Pitt had contemplated the measure, and it found many very able advocates in

England. O'Connell at first thought that the clergy should demand this arrangement; but, on their vehement opposition, he renounced the idea. In 1837 he had a warm controversy on the subjcet with Mr. Smith O'Brien, who advocated payment. Each was probably right, according to his own point of view. Mr. O'Brien looked mainly to the interests of his country—O'Connell to the interests of his Church. To pay the priests would have been, in a great measure, to pacify Ireland, but they would have been less powerful than when resting exclusively on the people, and they have always cared much more for power than for money.

On the accession of the Wellington Ministry to power the Catholic Association passed a resolution to the effect that they would oppose with their whole energy any Irish member who consented to accept office under it. When the Test and Corporation Acts were repealed, Lord John Russell advised the withdrawal of this resolution, and O'Connell, who, at that time, usually acted as moderator, was inclined to comply. Fortunately, however, his opinion was overruled. An opportunity for carrying the resolution into effect soon occurred. Mr. Fitzgerald, the member for Clare, accepted the office of President of the Board of Trade, and was consequently obliged to go to his constituents for re-election. An attempt was made to induce a Major Macnamara to oppose him, but it failed at the last moment, and then O'Connell adopted the bold resolution of standing himself. The excitement at this announcement rose at once to fever height. It extended over every part of Ireland, and penetrated every class of society. The whole mass of the Roman Catholics prepared to support him, and the vast system of organisation which he had framed acted effectually in every direction. He went down to the field of battle, accompanied by Sheil, by the well-known controversialist Father Maguire, and by Steele and O'Gorman Mahon, two very ardent but eccentric Repealers, who proposed and seconded him. Mr. Steele began operations by offering to fight a duel with any landlord who was aggrieved at the interference with his tenants—a characteristic but judicious proceeding, which greatly simplified the contest. O'Connell, Sheil, and Father Maguire flew over the country, haranguing the people. The priests addressed the parishioners with impassioned zeal from the altar; they called on them, as they valued their immortal souls, as they would avoid the doom of the apostate and the renegade, to stand firm to the banner of their faith. Robed in the sacred vestments, and bearing aloft the image of God, they passed from rank to rank, stimulating the apathetic, encouraging the fainthearted, and imprecating curses on the recreant. They breathed the martyr-spirit into their people, and persuaded them that their cause was as sacred as that of the early Christians. They opposed the spell of religion to the spell of feudalism—the traditions of the chapel to the traditions of the hall.

The landlords, on the other hand, were equally resolute. They were indignant at a body of men who had no connection with the county presuming to dictate to their tenants. They protested vehemently against the introduction of spiritual influence into a political election, and against the ingratitude manifested towards a tried and upright member. Mr. Fitzgerald had always been a supporter of the Catholic cause. He was an accomplished speaker, a man of unquestioned integrity, and of most fascinating and polished manners. His father—who was at this time lying on his death-bed—had been one of those members of the Irish Parliament who had resisted all the offers and all the persuasions of the Ministry, and had recorded their votes against the Union. The landlords were to a man in his favour. Sir Edward O'Brien, the father of Mr. Smith O'Brien, and the leading landlord, proposed him, and almost all the men of weight and reputation in the county surrounded him on the hustings. Nor did he prove unworthy of the contest. His speech was a model of good taste, of popular reasoning, and of touching appeal. He recounted his services and the services of his father; and, as he touched with delicate pathos on this latter subject, his voice faltered and his countenance betrayed so genuine an emotion that a kindred feeling passed through all his hearers, and he closed his speech amidst almost unanimous applause. The effect was, however, soon counteracted by O'Connell, who exerted himself to the utmost on the occasion, and withheld no invective and no sarcasm that could subserve his cause. After two or three days' polling the victory was decided, and Mr. Fitzgerald withdrew from the contest.

Ireland was now on the very verge of revolution. The whole mass of the people had been organised like a regular army, and taught to act with the most perfect unanimity. Adopting a suggestion of Sheil, they were accustomed to assemble in every part of the country on the same day, and scarcely an adult Catholic abstained from the movement. In 1828 it was computed that in a single day two thousand meetings were held. In the same year Lord Anglesey had written to Sir Robert Peel, stating that the priests were working most effectually on the Catholics of th army, that it was reported that many of these were ill-disposed, and that it was important to remove the depots of recruits, and supply their place by English or Scotch men. The contagion of the movement had thoroughly infected the whole population. If concession had not been made, almost every Catholic county would have followed the example of Clare; and the Ministers, feeling further resistance to be hopeless, brought in the Emancipation Bill, confessedly because to withhold it would be to kindle a rebellion that would extend over the length and breadth of the land.

It was thus that this great victory was won by the genius of a single man, who had entered on the contest without any advantage of rank, or wealth, or influence, who had maintained it from no prouder eminence than the platform of the demagogue, and who terminated it without the effusion of a single drop of blood. All the eloquence of Grattan and of Plunket, all the influence of Pitt and of Canning, had proved ineffectual. Toryism had evoked the spirit of religious intolerance. The pulpits of England resounded with denunciations; the Evangelical movement had roused the fierce passions of Puritanism; yet every obstacle succumbed before the energy of this untitled lawyer. The most eminent advocates of emancipation had almost all fallen away from and disavowed him. He had devised the organisation that gave such weight to public opinion; he had created the enthusiasm that inspired it; he had applied to political affairs the priestly influence that consecrated it. With the exception of Sheil, no man of commanding talent shared his labours, and Sheil was conspicuous only as a rhetorician. He gained this victory not by stimulating the courage or increasing the number of the advocates of the measure in Parliament, but by creating another system of government in Ireland, which overawed all his opponents. He gained it at a time when his bitterest enemies held the reins of power, and when they were guided by the most successful statesman of his generation, and by one of the most stubborn wills that ever directed the affairs of the nation. If he had never arisen, emancipation would doubtless have been at length conceded, but it would have been conceded as a boon granted by a superior to an inferior class, and it would have been accompanied and qualified by the veto. It was the glory of O'Connell that his Church entered into the constitution triumphant and unshackled—an object of fear and not of contempt —a power that could visibly affect the policy of the empire.

The Relief Bill of 1829 marks a great social revolution in Ireland—the substitution of the priests for the landlords as the leaders of the people. For a long time a kind of feudal system had existed, under which the people were drawn in the closest manner to the landlords. In estimating the character of this latter class we must, I think, make very large allowance for the singularly unfavourable circumstances under which they had long been placed. The Irish Parliament was governed chiefly by corruption, and as the landlords controlled most of the votes, and as the county dignities to which they aspired were all in the gift of the Government, they were, beyond all other classes, exposed to temptation. They were also subject to much the same kind of demoralising process as that which in slave countries invariably degrades the slave-owner. The estate of the Protestant landowner had in very many cases been torn by violence from its former possessors. He held it by the tenure and in the spirit of a conqueror. His tenants were of a conquered race, of a despised religion, speaking another language, denuded of all political rights, sunk in abject ignorance and poverty, and with no leader under whom they could rally. Surrounded with helots depending absolutely on his will, it was not surprising that he contracted the vices of a despot. Arthur Young concludes a vivid description of the relation between the classes by the assertion that ' a landlord in Ireland can scarcely invent an order which a servant, or labourer, or cottier dares to refuse to execute ; ' and the total absence of independence on the part of the lower orders, and the general tolerance of brutal violence on the part of the higher orders, struck most Englishmen in Ireland. Besides this, the penal laws which gave the whole estate of the Catholic to any son who would consent to abjure his religion, seemed ingeniously contrived to secure a perpetual influx of unprincipled men into the landlord class; while the vast smuggling trade which necessarily followed the arbitrary and ruinous prohibition of the export of wool, conspired with other causes to make the landlords, like all other Irishmen, hostile to the law. The glimpses which are given incidentally of their mode of life by Swift, Berkeley, Chesterfield, and Dobbs, and at a later period by Arthur Young, are in many respects exceedingly unfavourable. The point of honour in Ireland has always been rather in favour of improvidence than of economy. In dress and living a scale of reckless expenditure was common, which impelled the landlords to rackrents and invasions of the common land, and these in their turn produced the agrarian troubles of the ' Whiteboys' and ' Hearts of Steel.' Hard drinking was carried to a much greater extent than in England, and both Berkeley and Chesterfield have noticed the extraordinary consumption of

French wines, even in families of very moderate means. The character of the whole landed interest is always profoundly influenced by that of its natural leaders, the aristocracy and the magistracy; but in Ireland peerages were systematically conferred as a means of corruption, and the appointments to the magistracy were so essentially political that even in the present century landlords have been refused the dignity because they were favourable to Catholic emancipation.[1] A spirit of reckless place-hunting and jobbing was very prevalent, and combined curiously with that extreme lawlessness which was the characteristic of every section of Irish society. Duelling was almost universal, and it was carried largely into politics, and even into the administration of justice; for a magistrate who gave a decision in favour of a tenant against his landlord was liable to be called out, and by the same process landlords are said to have defended their own tenants against prosecution. No Irish jury, Arthur Young assures us, would in duelling cases find a verdict against the homicide. It was a common boast that there were whole districts in which the King's writ was inoperative. In the early part of the eighteenth century 'hell-fire clubs,' which were scenes of gross vice, existed in Dublin, and the crime of forcible abduction was, through nearly the whole of the eighteenth century, probably more common in Ireland than in any other European country, and it prevailed both among the gentry and among the peasants. It is worthy of notice that Arthur Young observed in the

[1] The reader may find some very curious facts about the appointments of Irish magistrates in the early part of this century in O'Flanagan's 'Lives of the Irish Chancellors' (Life of Lord Manners); Lord Cloncurry's 'Personal Recollections;' and Bulwer's 'Life of Lord Palmerston,' vol. i. p. 337.

former, as much as in the latter, a strong disposition to screen criminals from justice.

These are the shades of the picture, and they are sufficiently dark. On the other hand, as the eighteenth century advanced, the character of the higher classes improved. Drinking and duelling, though still very general, had appreciably diminished. The demoralising influence of the penal laws was mitigated. The gentry were gradually rooted to the soil, and a strong national feeling having arisen, they ceased to look upon themselves as aliens or conquerors. The Irish character is naturally intensely aristocratic; and when gross oppression was not perpetrated, the Irish landlords were, I imagine, on the whole very popular, and the rude, good-humoured despotism which they wielded was cordially accepted. Their extravagance, their lavish hospitality, their reckless courage, their keen sporting tastes, won the hearts of their people, and the feudal sentiment that the landlord should command the votes of his tenants was universal and unquestioned. The measure of 1793, conferring votes on the Catholics, though it is said to have weakened the zeal of some of the advocates of Parliamentary reform, left this feeling unchanged. Nor were the Irish gentry without qualities of a high order. The love of witty society; the passion for the drama and especially for private theatricals, which was very general in Ireland through the eighteenth century; and, above all, the great school of Parliamentary eloquence in Dublin, indicated and fostered tastes very different from those of mere illiterate country squires. The noble efflorescence of political and oratorical genius among Irishmen in the last quarter of the century, the perfect calm with which great measures for the relief of the Catholics which would have been impossible in England were received in Ireland; above all, the manner in which the Volunteer movement was organised, directed, and controlled, are decisive proofs that the upper classes possessed many high and commanding qualities, and enjoyed in a very large measure the confidence of their inferiors. They were probably less uncultivated, and they were certainly much less bigoted, than the corresponding class in England, and as long as they consented to be frankly Irish, their people readily followed them. Occasional instances of deliberate tyranny and much sudden violence undoubtedly took place; but it should be remembered that during the whole of the eighteenth century the greater part of Ireland was let at very long leases, and that the margin between the profits of the tenant and the rent of the landlord was so great that the former almost invariably sublet his tenancy at an increased rent. The distress of the people was much more due to this system of middlemen, and to their own ignorance and improvidence, than to landlord tyranny; and the faults of the upper classes, in dealing with their tenants, were rather those of laxity and imprudence than of harshness. The absence of any legal provision for the poor produced great misery, and had a bad economical effect in removing one of the great inducements to the gentry to check pauperism; but, on the other hand, it fostered a very unusual spirit of private charity through the country. Absenteeism was much complained of; but this probably sprang more from the great tracts of confiscated land which had been given to great English proprietors, than from the systematic absence of the natives. The presence of a Parliament secured a brilliant society in Dublin; and in the country travellers represent the roads as rather better than in England, and the country seats as numerous and imposing. The absence of rival authority and of religious intolerance, and the character of the people, made the social system work better than might have been expected. Good-nature is, perhaps, the most characteristic Irish virtue; and if it is not one of the highest, it is at least one of the most useful qualities that a nation can possess. It will soften the burden of the most oppressive laws and of the most abject poverty, and the only evil before which it is powerless is sectarian zeal. O'Connell evoked that zeal, and the bond between landlord and tenant was broken. 'I have polled all the gentry, and all the 50*l.* freeholders,' wrote Mr. Fitzgerald to Sir R. Peel, when giving an account of his defeat—' the gentry to a man.' The attitude which the landlord class afterwards assumed during the agitation for Repeal completed the change, and they have never regained their old position.

It must be added that another important train of causes was operating in the same direction. The economical condition of Ireland had long been profoundly diseased. The effect of the confiscations, and of the penal laws, had been that almost all the land belonged to Protestants, while the tenants were chiefly Catholics. The effect of the restrictions on trade had been that manufacturing industry was almost unknown, and the whole impoverished population was thrown for subsistence upon the soil. At the same time the English land laws, which are chiefly intended to impede the free circulation and the division of land, were in force in the country in which, beyond all others, such circulation is desirable. One of the most important objects of a wise legislation is to soften the antagonism between landlord and tenant by interweaving their interests, by facilitating the creation of a small yeoman class who break the social disparity, and by providing outlets for the surplus agricultural population. In Ireland none of these mitigations existed; and the difference of religion, and the memory of ancient violence, aggravated to the utmost the hostility. The tithes, levied for the most part on the poor Catholics for the support of the Church of the landlords, were another element of dissension. All the materials of the most dangerous social war thus existed, though the personal popularity of the landlords, and the prostrate condition of the Catholics, for a time postponed the evil. The habits of disorder, and the secret organisations which had arisen in the middle of the eighteenth century, continued to smoulder among the people, and in the great distress that followed the sudden fall of prices which accompanied the peace, they broke out afresh. The land, as I have said, in the closing years of the eighteenth century, was chiefly let at moderate rents on long leases. The tenant usually sublet his tenancy, and on the great rise of prices resulting from the war, the sub-tenant usually took a similar course, and the same process continued till there were often four or five persons between the landlord and the cultivator of the soil. The peasants, accustomed to the lowest standard of comfort, and e couraged by their priests to marry early, multiplied recklessly. The land was divided into infinitesimal farms, and all classes seemed to assume that war prices would be perpetual. Many landlords, bound by their leases, were unable to interfere with the process of division, while others acquiesced in it through laxity of temper or dread of unpopularity; and others encouraged it, as the multiplication of 40s. freeholders increased the number of voters whom they could control. In such a condition of affairs, the fall in the value of agricultural produce after the peace proved a crushing calamity. Large sections of the people were on the verge of starvation, and among all agricultural labourers there was a distress and a feeling of oppression which alienated them from their landlords, and predisposed them to follow new leaders.

' When introducing the Roman Catholics to Parliament, the Ministers brought forward two or three measures with the object of diminishing their power, the only one of any real value being the disfranchisement of the 40s. freeholders. This measure greatly lessened the proportion of the Roman Catholic electors. It struck off a number of voters who were far too ignorant to form independent opinions, and it in some degree checked the fatal tendency to subdivision of lands. It would have been well if the Ministers had stopped here; but, with an infatuation that seems scarcely credible, they proceeded in this most critical moment to adopt a policy which had the effect of irritating the Roman Catholics to the utmost, without in any degree diminishing their power, and of completely preventing the pacific effects that concession might naturally have had. Their first act was to refuse to admit O'Connell into Parliament without re-election, on the ground that the Emancipation Act had passed since his election. It was felt that this refusal was purely political, and designed to mark their reprobation of his career. It was, of course, utterly impotent, for O'Connell was at once re-elected; but it was accepted by the whole people as an insult and a defiance. O'Connell himself was extremely irritated, and to the end of his life his antipathy to Sir Robert Peel was of the bitterest and most personal character. He said of him that 'his smile was like the silver plate on a coffin.' There was, perhaps, no single measure that did so much to foster the feeling of discontent in Ireland as this paltry and irrational proceeding.

It was succeeded by another indication of the same spirit. By the Emancipation Act the higher positions in the Bar were thrown open, as well as the Parliament. A distribution of silk gowns naturally followed; and, while several Roman Catholic barristers obtained this distinction, O'Connell, who occupied the very foremost position, was passed over. Among those who were promoted was Sheil, who had co-operated with him through the whole struggle. It now, too, became manifest that the Tories were determined to render the Emancipation Act as nugatory as was possible, by never promoting a Roman Catholic to the bench. For some time under their rule the exclusion was absolute. The Relief Bill was also accompanied by a temporary Act suppressing the Catholic Association, and enabling the Lord-Lieutenant, during the space of rather more than a year, to suppress arbitrarily, by proclamation, any association or assembly he might deem dangerous. A measure of this kind suspended every vestige of political liberty, and left the people as discontented as ever. O'Connell declared that justice to Ireland was not to be obtained from an English Parliament, and the tide of popular feeling set in with irresistible force towards Repeal. Of all possible measures, Catholic emancipation might, if judiciously carried, have been most efficacious in allaying agitation, and making Ireland permanently loyal. Had it been carried in 1795—as it undoubtedly would have been if Pitt had not recalled Lord Fitzwilliam—the country would have been spared the Rebellion of 1798, and all classes might have rallied cordially round the Irish Parliament. Had it been carried at, or immediately after, the Union—as it would have been if Pitt had not again betrayed the cause—it might have assuaged the bitterness which that measure caused, and produced a cordial amalgamation of the two nations. It was delayed until sectarian feeling on both sides, and in both countries, had acquired an enduring intensity, and it was at last conceded in a manner that produced no gratitude, and was the strongest incentive to further agitation. In estimating the political character of Sir R. Peel, it must never be forgotten that on the most momentous question of his time he was for many years the obstinate opponent of a measure which is now almost universally admitted to have been not only just, but inevitable; that his policy having driven Ireland to the verge of civil war, he yielded the boon he had refused simply to a menace of force; and that he accompanied the concession by a display of petty and impotent spite which deprived it of half its utility and of all its grace.

The exasperation of O'Connell at these measures was extreme. He denounced the Ministry of Wellington and Peel with reckless violence, endeavoured in 1830 to embarrass it by a mischievous letter recommending a run upon gold, revived the Catholic Association under new names and forms, and energetically agitated for the repeal of the Union. The proclamations of the Lord-Lieutenant, however, suppressed these associations, and when he attempted to hold public meetings he was compelled to yield to a prosecution; the upper classes strongly discouraged the new agitation, and the Ministry of Wellington soon tottered to its fall. In the beginning of 1831 he accordingly desisted from agitation, ostensibly in order to test the effect of emancipation upon the policy of the Imperial Parliament. The Reform question was at this time rising to its height. O'Connell advocated the most extreme Radical views, and, in 1830, brought in a Bill for universal suffrage, triennial Parliaments, and the ballot. He wrote a series of letters on the question. He brought the whole force of his influence to act upon it, and his followers contributed largely to the triumph of the measure of 1832—a fact which was remembered with great bitterness when the Reformed Parliament began its career by an extremely stringent Coercion Bill for Ireland.

The social condition of Ireland was, indeed, at this time most deplorable. Agrarian murders and tithe riots, the burning of houses and the mutilation of cattle, were of almost daily occurrence. Secret societies ramified over the country, and in a considerable part of Leinster absolute anarchy was reigning. The bonds that united society were broken, law was utterly discredited, and class warfare and religious animosity were supreme.

To a certain extent O'Connell was, undoubtedly responsible for these crimes. He had first awakened the Catholics out of their torpor, made them sensible of their wrongs, and taught them to look to themselves for the remedy. He had begun a fierce political agitation which propagated itself in various forms through all classes of the community. He had broken down the reverence for rank, set class against class, lashed an excitable people to frenzy by the most inflammatory language, distinctly encouraged them to refuse the payment of tithes, and palliated, or more than palliated, all the violence to which that refusal led. On the other hand, he uniformly denounced secret societies with unqualified severity, and represented them as the most fatal obstacles to his policy. 'He who commits a crime adds strength to the enemy,' was one of his favourite mottos, and he had few greater difficulties to encounter than the Coercion Bills which these lawless outbursts provoked. It should also not be forgotten, in considering the connection between political agitation and crimes of violence, that the latter almost disappeared in Ireland during the Repeal movement, when the former was at its height.

Whatever opinion, however, may be formed about the manner in which the blame of these outrages should be distributed, they are in themselves at least sufficiently explicable. A people, poor, ignorant, and extremely excitable, had been urged into a furious and most successful agitation. A fierce war of classes and a fierce religious animosity were raging, and at the same time the whole administration of justice and the whole local government were in the hands of men in whom the great majority of the population could have no confidence. In 1833—four years after Catholic emancipation—there was not in Ireland a single Catholic judge or stipendiary magistrate. All the high sheriffs with one exception, the overwhelming majority of the unpaid magistrates and of the grand jurors, the five inspectors-general, and the thirty-two sub-inspectors of police, were Protestant. The chief towns were in the hands of narrow, corrupt, and, for the most part, intensely bigoted corporations. Even in a Whig Government, not a single Irishman had a seat in the Cabinet, and the Irish Secretary was Mr. Stanley, whose imperious manners and unbridled temper had made him intensely hated. For many years promotion had been steadily withheld from those who advocated Catholic emancipation, and the majority of the people thus found their bitterest enemies in the foremost places. Their minds were now turned eagerly towards Repeal, and they were told by the English Minister that the constitutional expression of their desire would be perfectly useless, and that 'the people of England would resist it to the death.' At the same time, it is scarcely an exaggeration to say that the

British constitution had no existence in Ireland. Sir R. Peel, in one of his speeches in 1829, made an admission which is an instructive comment on the common eulogies of the pacifying wisdom of the Irish policy of Pitt. He stated that in scarcely one year since the Union was Ireland governed by ordinary law.[1] The Habeas Corpus Act, which is perhaps the most important part of the British constitution, was suspended in Ireland in 1800, from 1802 till 1805, from 1807 till 1810, in 1814, from 1822 till 1824.[2] There was no public provision for the poor. There was no system of national education except the Sectarian Kildare Street Society. Above all, while the Catholic priests received no payment from Government, the poorest Catholic cottager was compelled to pay something to support the hostile and aggressive Church of the rich minority. There are few methods of levying money which have been in general more unpopular than tithes, this impost being, as Paley observed, 'not only a tax on industry, but the industry that feeds mankind,' and of course the natural objections to it were immeasurably intensified when it was levied from a half-starving peasantry, who derived no religious benefit from the ministrations of those they were compelled to pay. A second rent, raised from the most impoverished classes of the community in favour of men who contributed nothing to production, and in order that they might oppose the religious convictions of those who paid them, was a grievance so monstrous, so palpable, and so universally felt, that it could not fail, when the Catholics acquired some measure of self-confidence, to produce a general conflagration. In the

---

[1] See Doubleday's 'Life of Sir R. Peel,' vol. i. pp. 482, 483.
[2] Sir E. May's 'Constitutional History,' vol. ii. p. 270.

eighteenth century the Whiteboys had been chiefly organised in opposition to the tithes, and the landlords were said sometimes to have instigated them. A resolution of the House of Commons in 1735, which was converted into a regular law just before the Union, relieved pasture in a great measure of tithes, thus throwing the burden chiefly on the cottier class; and there were some curious inequalities which Grattan exposed and denounced in the burdens imposed on the different counties. The clergy, by their profession and habits, were of course very unfitted to collect the tithes, and the extreme minuteness of Irish tenancies added greatly to the difficulty. Shortly before the tithes in Ireland were commuted it was stated officially that in a single parish in Carlow the sum owed by 222 defaulters was one farthing each, and that a very large proportion of the defaulters throughout the country were for sums not exceeding one shilling. Under these circumstances, the clergy very naturally farmed out their interest to tithe-proctors, who often exercised their rights with extreme harshness, and became more hated than any other class in the country. Grattan had vainly laboured to have tithes commuted, and much ecclesiastical superstition was shown in defending a system which, on grounds of expediency and grounds of equity, was utterly untenable. At last a general conspiracy to refuse payment spread over Ireland, and every kind of outrage was directed both against those who collected and those who paid them. The law was utterly paralysed. The clergy, deprived of their lawful income, were thrown into the deepest distress. Government came to their assistance by advancing 60,000*l.* in 1832 for the clergy who had been unable to collect their tithes in the preceding year, and it undertook to collect the unpaid tithes of

1831. The attempt was a signal failure. The arrears for that year were 104,000*l.*, and of that sum, after fierce conflicts and much bloodshed, the Government recovered 12,000*l.* at a cost of 15,000*l*. Scarcely anyone ventured to defy the popular will by paying the tithes. It was with difficulty that the ordinary legal process of distraint was executed; and when in obedience to the law the cattle or crops of the defaulter were put up to auction no one dared to buy them. A lawless combination, sustained by the consciousness of a real grievance, completely triumphed, and the presence of a Protestant clergyman was often sufficient to demoralise an entire district.

As I am not writing a history of Ireland, I shall only advert very briefly to the important measures by which the reformed Parliament endeavoured to check these evils. The first measure, as I have said, was a coercive Bill surpassing in stringency any to which Ireland had yet been made subject, and directed not only against crime, but also against political agitation. Among other provisions, it replaced the ordinary tribunals in the proclaimed districts by martial law; and it took away over the whole of Ireland all liberty of political meeting and discussion. That some measure of severe coercion was necessary is incontestable; for it was computed that in 1832 there were more than 9,000 crimes perpetrated in Ireland connected with the disturbed state of the country, and among them nearly 200 cases of homicide. At the same time martial law, which was equivalent to a total suspension of the constitution, was a measure of extraordinary though perhaps not excessive severity, and appeared especially so in Ireland, where the atrocities perpetrated under that law in 1798 were still remembered. The part, however, of the Coercion Bill which excited the most intense and most natural animosity was that which was directed against political action. The repeal of the Union, whether it was wise or the reverse, was an object at which it was perfectly constitutional to aim. Parliament had an undoubted right to effect it, and therefore the people had an equally undoubted right to petition for it. If it had been constitutional before 1800 to advocate a union, it was equally constitutional after 1800 to advocate its repeal. Agrarian and tithe outrages were chiefly reigning in one of the four provinces; but by the Coercion Bill of Mr. Stanley no political meeting could be held in any part of Ireland without the express permission of the Lord-Lieutenant. The King's speech, which foreshadowed the measure, like two preceding ones, contained a paragraph directed against O'Connell and his agitation, and the Coercion Bill appeared especially obnoxious, as coming from a Whig Ministry in a reformed Parliament, immediately after the Reform Bill which O'Connell had contributed not a little to carry.

It is scarcely possible, without possessing the detailed evidence which is at the disposal of a government, to pronounce with confidence upon whether the state of the country required or justified these clauses. It is not, however, surprising that they exasperated O'Connell to the highest degree; and at no period of his career was his language more violent than during the Ministry of Lord Grey. It was at this time that he talked of the 'base, bloody, and brutal Whigs,' and described them as men 'with brains of lead and hearts of stone and fangs of iron.' He and Mr. Stanley hated one another with the most intense hatred; and Parliamentary oratory contains very few instances of fiercer and more powerful invective than they exchanged. Sir Robert Peel and the Tories strongly supported the Coercion Bill, and the House was generally bitterly hostile to O'Connell; but the extraordinary vigour and eloquence of his opposition had at length their reward. The Coercion Bill was carried in 1833, but a strong feeling against its political clauses was aroused among Liberals; and when it was intended to renew them in the following year, there was a dissension in the Cabinet, of which O'Connell was informed, and which he disclosed in the House. The result was that the Coercion Bill was only re-enacted in a modified form and without the political clauses. Lord Grey retired from office, and Lord Melbourne became the head of a Ministry of which O'Connell was the chief support.

The measures, however, which were carried by the reformed Parliament were not simply coercive; they were also in a very large measure remedial. The subject which, if not the most important, was at least the most eagerly discussed, was the Irish Church; and there was none upon which O'Connell felt more keenly. Himself a fervent Catholic, the main object of his policy was to raise the Catholics out of the condition of a proscribed and degraded caste; and there is much reason for believing that he would have given up the notion of Repeal if he could have otherwise secured this equality. With the exception of his advocacy of Repeal, no part of his Irish policy injured him so much in the eyes of the English people as the opinions he hazarded about the Church; but, judged by the light of the events of our own day, they will be pronounced very reasonable and very moderate. He never appears to have advocated the withdrawal of all revenue from the Protestants, nor did he desire any further assistance than glebes to be given to the priests. The details of his proposal were more than once varied, but the main object was to put an end to the grievance of tithes. The Church lands he was willing to leave wholly or in a very great degree with their present possessors, and they would furnish a revenue which with very moderate assistance from voluntary sources would be amply sufficient for the real wants of the Protestants. The tithe fund before all things was to cease to be a tribute paid to the Protestant Church. About its disposition there was much difference of opinion. Probably the most popular solution would have been the simple cessation of the tithe payment, and this would have been a benefit both to the landlords and tenants; but other schemes, such as applying the fund to secular instruction or to building new charitable institutions, were advocated; and O'Connell appears finally to have settled upon the precise disposition which many years after his death was adopted by Mr. Gladstone. 'My plan,' he said, in a letter to Mr. Sharman Crawford in September 1834, 'is to apply the fund in the various counties of Ireland to relieve the occupiers of land from grand jury cess, ... to defray all the expenses of dispensaries, infirmaries, hospitals, and asylums, and to multiply the number of these institutions until they become quite sufficient for the wants of the sick.'

In this, however, as on many other points, O'Connell was considerably in advance of his age. With the exception of a few Radicals, no class in England would have tolerated such a measure. A growing school at Oxford and in the country looked upon all interference with Church revenues as sacrilege, and the famous work of Mr. Gladstone embodied and widely diffused what may be called the transcendental arguments in favour of establishments. Sir R. Peel admitted that the State had a right to change and regulate the distribution of Church revenues, but he denied that it had any right to divert them from Church purposes; and, in the case of the Irish Church, he maintained on the ground of the Act of Union, that disendowment would be a distinct breach of faith. That Act, he said, 'differs in this respect from an ordinary law, that it was a national compact, involving the conditions on which the Protestant Parliament of Ireland resigned its independent existence. In that compact express provision is made which, if anything can have, has an obligation more binding than that of ordinary law. .... A right was reserved in that Act with respect to the removal of the civil disabilities of the Catholics, but no right was reserved to the United Parliament to deal with the property of the Church of Ireland.'

The Tory party, therefore, whether they adopted the extreme views of the new Oxford school or the more moderate views of Sir Robert Peel, were united in resisting any diminution of the revenues of the Church; and they could enlist in their cause the two cries of 'No Popery' and 'the Church in danger,' which were probably the most powerful in England. The Whigs were not equally united. A small but very able section agreed with Sir Robert Peel that the power of Parliament extended only to the redistribution, but not to the alienation of ecclesiastical revenues. The main body, however, including Lord Grey, Lord Althorp, and Lord John Russell, maintained that Parliament had a right, when the wants of the Protestants were adequately supplied, to apply the surplus revenues of the Church to purposes of education or of charity that would be beneficial to the whole community. The first attempt to carry out this policy was in the Ministry of Lord Grey, when a clause was introduced in the 'Church Temporalities Act,' to give Parliament the disposal of a surplus resulting from the grant of perpetual leases of Church lands; but this clause, which was very restricted in its operation, was abandoned in committee as likely to endanger the success of the Bill. The subject was once or twice renewed during the same Ministry, and the opinion of the Government clearly pronounced, but nothing decided was done till Sir Robert Peel came into office. He was governing with a minority of the House, and his Ministry was obviously ephemeral. He brought in a measure for commuting Irish tithes in 1835, when Lord John Russell moved as an amendment the famous Appropriation Clause, affirming that any surplus revenues of the Irish Church not required for the religious wants of the Protestants should be applied to the moral and religious education of the people at large, and that no measure concerning tithes would be satisfactory which did not embody this principle. The resolution was carried, Sir R. Peel retired from office, and Lord Melbourne became Prime Minister.

If it be considered as a mere party move, there has seldom been a more disastrous mistake than that of the Whigs in bringing forward this Appropriation Clause, and in selecting it as the question on which to overthrow the first feeble Ministry of Sir R. Peel. At the same time, there never was a more loyal or moderate attempt to remedy a great injustice. By the confession of all parties, the existing condition of the Church was scandalous in the extreme, the number and emoluments of the bishops were absurdly out of proportion to the numbers of their flocks, and there were 151 instances of parishes containing not a single Protestant. Few persons will now deny that the Church revenues might have been justly diminished, or that an application of a portion of them to the benefit of the whole community would have strengthened the position of the Church. The Ministry of Lord Melbourne, however, soon found the task they had undertaken beyond their powers. Lord J. Russell, as Minister, duly brought in the clause as a portion of the Bill for commuting tithes, but although it was carried through the House of Commons it was only by a small majority, and a majority of the English members were against the Government. Mr. Stanley, the most brilliant orator, and Sir J. Graham, who was one of the ablest administrators of the Whigs, with a few others, had seceded from the party on this question as early as 1834, and were strenuously opposed to the Government. O'Connell ridiculed the small number of the secessionists, quoting with great effect the lines of Canning—

> Adown thy dale, romantic Ashbourne, glides
> The Derby Dilly with just *six* insides.

But the ability and the political weight then withdrawn from the Whigs were never adequately replaced. The violence of O'Connell, who supported the Appropriation Clause with passionate zeal, produced a strong Conservative reaction in England. The King was known to be opposed to the policy of his Ministers, and the House of Lords by large majorities rejected the clause. In the meantime the tithes question continued in abeyance, and it was plain that until it was settled there could be no real peace in Ireland. There were not wanting those who urged the Ministers, as the sole means of carrying their Bill, to avail themselves of the fierce Radical spirit which was abroad, and which demanded the subversion of the House of Lords or its organic change. Happily, however, those who then guided the policy of England were deeply and fervently attached to the constitution. Had they persevered, a violent revolutionary spirit might have arisen; and, by abandoning the Appropriation Clause in 1838, they probably saved the country from an irretrievable disaster at the cost of a ruinous party humiliation.

But although this measure failed, two important Church reforms were carried. By 'the Church Temporalities Act of 1833,' the revenues of the Church were redistributed and its most excessive abuses corrected. Two archbishoprics and eight bishoprics, as well as a number of minor dignitaries, were abolished. Considerable reductions were made in the revenues of the other bishoprics, and provision was made out of the surplus thus obtained for augmenting small livings and building glebes and churches. The Establishment was thus made more defensible than before. If it continued to be an anomaly it ceased to be a scandal; its offices were no longer pampered sinecures, and its dignities at last bore a fair proportion to the number of its worshippers. In one respect the Bill was a benefit to the Catholics, for the Church cess, which had been levied chiefly from Catholics and dispensed by Protestant vestries, was replaced by a tax upon the clergy for the repair of churches. The unceremonious way in which superfluous bishoprics were abolished gave great offence in some quarters in England, and was one of the proximate causes of the Tractarian movement. A still more important reform was after long delay and many vicissitudes at last effected in 1838, with the concurrence of both of the great parties in Parliament. I mean the substitution of a land tax for the old system of tithes. By this substitution the burden was removed from the peasants, who were nearly all Roman Catholics, and imposed on the landlords, who were nearly all Protestants. Twenty-five per cent. was taken off the clerical income derived from tithes in consideration of the certainty, facility, and inexpensiveness of its collection under the new system.

This measure was violently opposed by O'Connell, who desired to see the tithes either simply abolished or diverted from Church purposes, a course which would undoubtedly have been the most popular in Ireland. It was contended by the political economists that the change would give no real relief to the tenant, as the burden that was transferred to the landlord would be met by a corresponding increase of rent. But this, like all similar doctrines of political economy, is true only in as far as land is dealt with simply and rigidly on commercial principles, and in Ireland as a matter of fact it has never generally been let at the extreme competitive price. Of this fact the great place which the middlemen occupy in Irish agrarian history is a decisive proof. The Irish landlords readily assumed the burden in consideration of the land tax being applied to the support of their own Church, and the rents were not, I believe, in general raised. It is worthy, too, of notice, that when the Established Church was recently disendowed, no voice outside of the landlord class was raised in favour of simply abolishing the land tax, although that tax was said to have been in reality paid by the occupying class, and although it is probable that the majority of that class, if they had been consulted in 1835, would have voted for the abolition of tithes.

The tithes composition measure had the disadvantage of being conceded, like most Irish measures, to violence, and it has not proved a final arrangement. Subject to these qualifications, however, it deserves the highest praise. Few laws have ever been so completely successful in eradicating a great source of crime and allaying dangerous agitation. The

Protestant clergy, constituting a class of country gentry where such a class was peculiarly needed, and discharging many charitable and civilising functions towards the Catholic population, have, when they have abstained from active proselytising, been in general eminently popular, and the signal devotion which they manifested amid the horrors of the famine obtained for them a large measure of well-earned gratitude. During the last twenty-five years, in the worst periods of Irish crime, and in the worst localities, they have invariably been unmolested and unmenaced. With the exceptions of the priests and of converts, no class of Irishmen has been very bitterly opposed to them, and probably few great measures have excited less genuine enthusiasm in Ireland than the English measure for disendowing them.

In addition to these measures, others of great importance were taken. The system of national education, like all the branches of Irish administration, had been for a long time grossly unjust towards the Catholics. The Charter schools of Primate Boulter were distinctly proselytising, and some of the most iniquitous of the penal laws were those which forbade Catholics from engaging in the work of education. The 'Kildare Street Society,' which received an endowment from Government, and directed national education from 1812 to 1831, was not proselytising, but its management fell into the hands of the Evangelical party, which was rapidly rising in Ireland, and a rule was adopted, making the reading of the Bible without note or comment compulsory in its schools. Such a rule was in direct contradiction to the teaching of the Catholic Church, and it naturally created very general discontent. In 1831, however, and 1832, a system of national education was founded in Ireland, which continues, though seriously modified, to the present day. It was chiefly devised by Lord Anglesey, Mr. Plunket, Mr. Stanley, Mr. Blake, and Lord Cloncurry, and was intended to give the whole mass of the people a united secular education, while it offered facilities for separate religious education. A large proportion, however, of the Protestant clergy discovered that there were sundry passages in the Old Testament and in the Ordination Service which made it criminal for them to take part in any system of education in which they were not allowed to teach all their pupils the Bible, and they accordingly set up a rival system, which still exists, and they thus threw the national education to a great extent into the hands of the priests. These latter, however, gradually became more and more Ultramontane; it became one of their great ends to prevent the members of the two religions associating, and to impregnate all teaching on purely secular subjects with their distinctive ecclesiastical tenets; and they accordingly grew very hostile to the National Board. The original system was much tampered with to meet their wishes. The Church Education schools, in which the Bible is taught to everyone, are still unassisted by the Government, but endowments have been freely given to sectarian convent schools managed by monks or nuns. But these unjust, because unequal, departures from the original design have not saved the national system from the unanimous condemnation of the Catholic hierarchy. On the whole, that system has conferred upon the rising generation of Irishmen the inestimable blessing of a sound secular education; it has contributed in some degree to allay the animosity of sects; and it would, I believe, be difficult to cite a single instance of a Catholic who has become a Protestant,

T

or a Protestant who has become a Catholic, under its influence.

The liberal educational policy of the Whigs was fully adopted and extended by the Tory Government of Sir R. Peel. The College of Maynooth, intended for the education of the Irish priests, is one of the few existing institutions which owe their origin to that Irish Parliament which is so often represented as the hotbed of bigotry. It was founded during the viceroyalty of Lord Fitzwilliam, when the French war excluded Irishmen from France, and when the dread of an influx of French ideas was very strong, and it was a great boon to the Catholics, who previously possessed no means of educating their own clergy in their own land. Sir R. Peel in 1845, besides granting 30,000*l.* for building and improvements, nearly tripled the annual grant, and gave it a character of permanence by charging it on the consolidated fund. He in the same year established the three Queen's Colleges, in which a perfectly unsectarian education was provided—an advantage of which, in spite of many priestly anathemas, the Catholics have largely availed themselves.

Two other measures completed the work of reform. Although Ireland was one of the poorest countries in Europe—although a very large proportion of its population were continually on the verge of starvation—no legal provision existed for the destitute until 1838, when the Irish Poor-law was enacted. Although the corporations had been legally thrown open to Catholics in 1793, their constitution was so close that the admission was practically illusory, and the principal cities of an essentially Catholic country were almost exclusively governed by Protestants. For forty-seven years after the Catholics had been made eligible not one was elected into the corporation of Dublin. To remedy this gross injustice, the Government of Lord Melbourne, having carried a measure reforming the English corporations, brought forward in 1835 a similar measure for Ireland, but it was ardently opposed by Sir Robert Peel, and rejected by the House of Lords. The Tory party was naturally alarmed at the transfer of power that would be effected, and O'Connell had injudiciously predicted that the corporations would be 'normal schools of agitation.' The House of Lords was willing to abolish the close corporations, but refused to appoint new bodies, and proposed to destroy all municipal government in Ireland, and to substitute for the municipal authorities functionaries appointed by the Crown. The contest between the two Houses was as obstinate as about the Appropriation Clause, and it continued till 1840, when it was ended by a compromise. The Bill was passed, but only in a curtailed and mutilated form, and fifty-eight corporations were abolished. O'Connell soon afterwards became Lord Mayor of Dublin—a triumph which occasioned among his followers much vulgar and paltry glorification, but which was really under the circumstances of some importance—and a petition in favour of Repeal was voted by a large majority of the corporation.

All these measures were the consequence of the new political importance which the Catholics had acquired, and of the pressure which they exerted upon public opinion under the influence of O'Connell. Considerable however as they were, they by no means satisfied the great agitator, who would be content with nothing short of a complete destruction of the edifice of ascendency, and who had strong special objections to two of the measures I have enumerated. As the mouthpiece of the priests, he denounced the Queen's

Colleges as 'godless colleges,' borrowing the phrase and adopting the argument of Sir R. Inglis, a leader of the most extreme type of Tory. His objections to the poor-laws were of a different kind. He maintained with great force of argument the most rigid and most unpopular doctrines of the economists concerning the evils of guaranteeing relief to able-bodied paupers, and he also argued that a legal provision for the poor would check the spontaneous charity for which the lower classes in Ireland were remarkable, and that the workhouses would prove dangerous to female purity. The extent and intensity of Irish poverty he had no disposition to underrate, but the remedies he proposed were of a different kind. He flung the whole weight of his influence into the temperance movement, and he urged upon the Government the propriety of abolishing tithes, imposing a tax upon absentees, and giving assistance to emigration, which he justly looked upon as the only remedy that was adequate to the disease. Sir R. Peel, on the other hand, maintained that the long sea voyage would always stand in the way of its adoption to any considerable extent.[1]

While maintaining these views on Irish politics, he adopted on imperial questions the programme of the most extreme Radicals, advocating manhood suffrage, vote by ballot, short Parliaments, and the substitution of an elective for an hereditary Upper House. This was perhaps the gravest error of his career, and the extravagance of his opinions, and the incendiary and vituperative language with which he defended them, alienated from him the great majority of educated men, and made his alliance with the Whigs a source of weakness to his friends. Nor does he, I conceive, in

---

[1] 'Annual Register,' 1837, p. 70. At the time of the famine, however, Peel recommended Government aid to emigration.

this part of his career, deserve much credit for sincerity. The levelling disposition, the envious hatred of superiority and rank, which characterises the genuine English Radical, was wholly foreign to his nature, and is indeed rarely found among Irishmen. His loyalty to the Sovereign was very warm, and not unfrequently showed itself in language of almost Oriental servility. His democratic crusade was probably simply an incident of his Irish policy. An Irishman and a Catholic above all things, passionately attached to his country and his creed, he attacked with but little scruple any institution which stood in their way. To make numbers rather than wealth the source of political power would be to increase the relative importance of Ireland in the Empire, and of the Catholics in Ireland. In judging his conduct, we must remember that his policy was chiefly opposed by the aristocratic part of the State, that the House of Lords had steadily and persistently defeated or mutilated every attempt to raise the Catholics into equality with the Protestants, that the bitterest invectives were continually directed against him within its walls, and that it appeared idle to expect that Irish tithes could ever be abolished with its consent. Lord Lyndhurst pronounced the Irish to be 'aliens in race, in country, and religion.' O'Connell retorted by fierce denunciations of an hereditary caste overriding for selfish purposes the decisions of the representatives of the people. The Tory party desired to restrict the franchise in Ireland; they had already abolished the forty-shilling freeholders, and Lord Stanley long afterwards [1] attempted to carry the same policy still farther by imposing a system of registration so cumbrous and so troublesome that, if it had not been defeated by the Whigs, it would have virtually dis-

[1] In 1840.

franchised multitudes. O'Connell met this policy by maintaining the natural right of every man to a vote. His opponents in England appealed without the smallest scruple, and with eminent success, to the anti-Papal and anti-Irish feeling which was so strong in the lower strata of the English population. He retaliated by placing himself at the head of the wild movement for radical reform, and he carried his propagandism, not only into the great towns of the north of England, but also into Calvinistic Scotland. The party was at this time singularly deficient in eloquence, and Hume, who was its most influential member, was perhaps the most tangled and inarticulate speaker who ever succeeded as a leader in England. 'He would speak better,' O'Connell once said, 'if he finished one sentence before he began the next but one after.' O'Connell, trusting to his marvellous powers of popular oratory, defied religious prejudices and national antipathy, and rarely failed to win a momentary triumph; but his language was not suited to a cultivated English taste, and the revolutionary opinions he advocated, and the coarse personal abuse in which he continually indulged, justly lowered his influence with all sober persons.

The part which he played in imperial politics was, however, far from contemptible. Perhaps the three most important Parliamentary measures of the present century are the emancipation of the Catholics, the Reform Bill of 1832, and the establishment of free trade in corn. The first was chiefly due to O'Connell. In one of the most important divisions in the first Parliament of William IV. his followers turned the balance in favour of the second. He was an early and a strenuous advocate of the third. Unlike those petty traitors who, while professing to follow in his steps, have associated the cause of Irish nationality with the defence of negro slavery in America, of foreign military occupation in Italy, of Imperialism in France, and of assassination at home, he was steadily Liberal in every part of his policy. Parliamentary reform, free trade, the emancipation of negroes, the abolition of flogging in the army, the wrongs of Poland, the repeal of the taxes on knowledge, were among the causes he most ardently defended. Exercising an absolute authority over a large body of members, and availing himself with great skill of the divisions of parties, he was always a great power in the House of Commons, though he never succeeded in altogether catching its tone.

In debate he had to contend with almost overwhelming obstacles. All parties were in general combined against him, and all the great English speakers were his opponents. On Irish questions he had the immense disadvantage of speaking amid the derisive clamour of his audience, while his adversaries were cheered to the echo. The great majority of English and Scotch members regarded him with the strongest personal and political antipathy, and, with the exception of Sheil—who, though a very brilliant rhetorician, was scarcely a great debater—no Irish member was able to give him any considerable assistance. Almost all the eminent men he had to encounter had entered Parliament very young, and had attained their skill in debate and their knowledge of their audience by a Parliamentary education of many years. O'Connell came into the House of Commons at fifty-four; and a life spent in practising at the Irish Bar, or haranguing an Irish populace, was an exceedingly bad preparation for a Parliamentary career. But, notwithstanding all these disadvantages, his success as a debater was very great. His boundless readiness, his power of terse, nervous, Demosthenic reasoning, his thorough mastery of the subject he treated, the skill with which he condensed and pointed his case, and the rich flow of his humorous or pathetic eloquence, placed him at once in the foremost rank. At the same time, his speeches were extremely unequal. It would be easy to point out many that were masterpieces of masculine power, but yet they were continually defaced by coarseness and scurrility, by recklessness of assertion, and by extravagant violence. In a discussion on agricultural distress he scandalised all honest men by proposing as the sole adequate remedy a compulsory reduction of the national debt. He never obtained credit for a high sense of honour, and he was lamentably deficient in self-respect. The tact which he always manifested in dealing with the populace sometimes deserted him signally in an assembly of gentlemen, and, although none of his contemporaries could argue a particular question with a more commanding power, the general effect of his speaking upon the educated classes in England was certainly far from favourable. As a rhetorician he was surpassed by Sheil and Macaulay, but as a debater he was perhaps only equalled by Mr. Stanley, who, though probably greatly his inferior in general intellectual capacity, brought to the contest a far purer taste and a still fiercer temper, as well as a wonderful command of graceful and vigorous English, and an almost unrivalled dexterity in dialectic encounter.

The debates on Mr. Stanley's Coercion Bill were perhaps the most splendid examples of his Parliamentary powers. Assisted only by occasional speeches by Sheil, he had to bear the brunt of all the eloquence of Macaulay, Stanley, and Peel, together with numbers of minor orators, while Lord Brougham was inveighing against him in the other House. Notwithstanding these powerful odds, it seems to have been very generally admitted that in eloquence and in force he at least held his position throughout. O'Connell described the measure as a Bill directed against a single individual—himself. The interruptions he met with were sufficient to disconcert any less practised orator. On one occasion his voice was completely drowned for some time by an explosion of this inarticulate eloquence. When it had a little subsided, he exclaimed with characteristic impetuosity that he was not going to be put down 'by beastly bellowings;' upon which a member rose, and gravely observed that the epithet 'beastly' was out of order when applied to the exclamations of members of the House. O'Connell professed his willingness to retract the obnoxious expression, but added some apologetical remark to the effect that he had never heard of any bellowings that were not beastly. The Speaker decided that the epithet was contrary to order, but not more so than the ejaculations that elicited it.

His position towards English parties during a great part of his career was one of neutrality. The Tories he naturally detested, as the avowed enemies of Catholic emancipation and of reform; the Whigs he at one time defined as 'Tories out of place;' and there was no Ministry to which he was more hostile than that which originated the Coercion Bill. When, however, Lord Melbourne came into power, O'Connell gave his Ministry the whole weight of his support. His opponents Lord Grey and Mr. Stanley were no longer in the Ministry. The political clauses of the Coercion Bill had been abandoned. The Melbourne party had for the first time had the courage, by the appropriation clause, to attempt the application of some small parts of the Irish Church property to purposes of general utility, and the Irish Administratration of Lord Mulgrave, Lord Morpeth, and Mr. Drummond was eminently liberal and just. The Melbourne Ministry exhibited the rare spectacle of a Government opposed by the majority of the English members in the House of Commons, and by the great majority of the House of Lords, and at the same time unpopular with the country, but kept in power by the votes of the Irish members. O'Connell supported it very loyally, and although in his position there was perhaps no great merit in not being a place-hunter, it is worthy of notice how cheerfully he acquiesced in his exclusion from a Ministry of which he was for some time the mainstay. On questions of persons and offices the Ministers found him uniformly moderate and conciliatory, and in this respect his attitude formed a marked contrast to that of Lord Brougham. In 1838 he refused one of the highest legal positions in Ireland —that of Chief Baron. The Repeal cry at this time was suffered to sink, and in Ireland as in England O'Connell steadily and powerfully supported the Ministry.

There can, however, be no question that his support was ultimately a source of weakness. O'Connell was the especial bugbear of the English people—as he himself said, 'the best-abused man alive.' As the typical Irishman, Catholic, and Repealer, he aroused against himself the fiercest national and religious prejudices of large classes of Englishmen, while others were scandalised by his violent agitation for democratic reform, by his advocacy of free trade in corn, and by the coarse, reckless, and vituperative language in which he continually indulged. The downfall of the Melbourne Ministry and the complete triumph of Sir R. Peel were due to many causes which it is not within the object of the present work to investigate, but among them the almost universal dislike of O'Connell in England, and the undoubted fact that the Ministry subsisted mainly by his support, were prominent. The appropriation clause led to a great party humiliation, because it was plainly repugnant to the wishes of the great majority of the English people, and the anti-Popery and anti-Irish feelings were chief elements of the strong popular sentiment against the Government. It would have been impossible to give O'Connell a place in it without shattering it, and there was no taunt against Ministers more applauded than their alleged subserviency to the agitator. The House of Commons seldom rang with more enthusiastic plaudits than when Mr. Stanley, in one of his attacks upon the Government, quoted these lines from Shakespeare:

> But shall it be that you, that set the crown
> Upon the head of this forgetful man,
> And, for his sake, wear the detested blot
> Of murd'rous subornation—shall it be
> That you a world of curses undergo,
> Being the agents, or base second means,
> The cords, the ladder, or the hangman rather?
> Oh! pardon me that I descend so low
> To show the line and the predicament,
> Wherein you range under this subtle king.
> Shall it, for shame, be spoken in these days,
> Or fill up chronicles in time to come,
> That men of your nobility and power
> Did 'gage them both in an unjust behalf,
> As both of you—God pardon it!—have done?
>   .   .   .   .   .   .
> And shall it, in more shame, be further spoken,
> That you are fooled, discarded, and shook off
> By him for whom these shames ye underwent?

With purely political classes the Repeal policy of O'Connell was the chief cause of his unpopularity.

English politicians of all classes were united against it, and the almost unanimous denunciation of the scheme by all sections of the English press has so discredited it in England that it is somewhat difficult to do justice to its supporters. In the opinion of the present writer Repeal in the form in which it was advocated by O'Connell would have been equally injurious to Ireland and to the empire; but there was more to be said for the agitators than is commonly admitted. It should be remembered that O'Connell was old enough to recollect that Irish Parliament which he desired to restore; that that Parliament, with all its faults, contained a greater amount of genius and patriotism than has ever, either before or since, been engaged in the administration of Irish affairs; and that, whatever may have been the opinion of English statesmen, the unbribed talent of Ireland was almost unanimous against the Union. Canning had wittily compared the project of restoring the Irish Parliament to a project for restoring the Heptarchy, but an Irish politician who knew that the Parliament had existed only thirty years before, that many of its members remained, and that all the local ties and associations it had formed were still full of life, might be pardoned for thinking the comparison more ingenious than just. It should be remembered too, in estimating the sincerity of O'Connell, that he had made his maiden speech against the Union; that he had declared in that speech that, so far from desiring to purchase emancipation by the Union, he would rather the whole penal code should be re-enacted than that the Union should be passed; that he had reverted again and again to the subject before emancipation had been carried; and that in his dissatisfaction with the Union and its results he probably reflected the judgment, or at least the feeling, of some five-sixths of the people of Ireland. He advocated Repeal partly, no doubt, on the broad ground of nationality, but much more frequently on account of most definite grievances. He repeatedly urged as his main reason that he could not obtain 'justice to Ireland' from the Parliament in Westminster, and by this phrase he appears to have meant a condition of government in which, in the eyes of the State, an Irish Catholic should be placed on a level of perfect equality with an English or an Irish Protestant. This aim has been of late years fully attained, but the very magnitude of the measures which Parliament has thought necessary to its accomplishment is a justification of the complaints of O'Connell. It was impossible that emancipation, conceded in the manner it was, should have been accepted by the Catholics as sufficient, and before the measure had been carried O'Connell, in evidence before Parliament, had frankly said 'that unless it was done heartily and cordially it would only give them an additional power, and leave them the stimulant for exerting it.' An attempt had been made to deprive Ireland of all municipal freedom, and directly or indirectly to disfranchise the great body of the Catholics. The Tory party had so disposed of Government appointments as virtually to continue the system of disqualification, and when the Melbourne Ministry endeavoured to act with equality between the two religions, and in some small degree to modify the position of the Protestant establishment, it was destroyed chiefly on this very ground, and by the force of the anti-Irish and anti-Catholic feeling in the country the Tory Minister was replaced in power. It was then, and then only, that O'Connell threw himself heart and soul into the struggle for Repeal.

That he was by nature an agitator cannot be denied;

and, while he was extremely ambitious, he was at no time very high-minded or very scrupulous. One of the most remarkable features of his character was the steady and laborious perseverance with which through years of difficulty and discouragement he could subordinate all his many-sided activity to a single ambitious aim. In 1798, when still a young and unknown lawyer, he went through a very dangerous illness, and it is related of him that when he believed himself to be dying he was heard repeating those fine lines in 'Douglas:'

> Unknown I die. No tongue shall speak of me:
> Some noble spirits, judging by themselves,
> May yet conjecture what I might have proved,
> And think life only wanting to my fame.

He was not one of those men who could ever, like Washington, have been content, when he had conferred one great blessing upon his countrymen, to retire in the full enjoyment of his faculties from the arena. He loved power and popularity too much. His energy was inexhaustible. He delighted in being continually in the mouths of men, and in exercising that power of swaying great crowds which is at once one of the most intoxicating and one of the most dangerous of human gifts. But when all this is admitted it remains true that he was much less the creator than the director of the Repeal agitation, and that during a great part of his career he acted rather the part of a moderator than of an incendiary. He allayed agitation during the short administration of Canning. He acquiesced with scarcely a show of reluctance in the necessary disfranchisement of the 40s. freeholders. If his conduct after the Relief Bill was very violent, the measures that accompanied it, the fierce spirit which a protracted struggle had aroused, and the danger of allowing such agitators as Feargus O'Connor to direct the storm, do much to palliate his violence. During the whole of the Melbourne administration he kept the question of Repeal in abeyance, and distinctly said that he would abandon the notion if the English Parliament would do justice to Ireland. In the second Ministry of Peel, it is true, he threw away the scabbard and threw all his energies into the struggle for Repeal; but even in 1843, in the very zenith of the movement, he wrote a letter giving a decided preference to the much more moderate scheme of a federal Union, under which the Irish Parliament should be restricted to local affairs, while an Imperial Parliament should manage imperial ones. The wishes of the people and the policy of the Tories in a great measure forced him into agitation, and he abandoned federalism simply because he found it almost universally unpopular.

These facts are sufficient to show that O'Connell was not the selfish and reckless incendiary he is sometimes considered. At the same time, in spite of considerable hesitation and timidity in action, he was ever at heart a fervent Repealer. Endowed to an extraordinary degree with that 'retrospective imagination' which is so characteristic of his countrymen, the recollection of the Irish Parliament had been the earliest romance of his life. His ambition had been first kindled by those orators who shed a glow of such immortal eloquence over its fall. In him as in many Irishmen the shameful story of the Union awoke passions of the bitterest and most enduring resentment, and the possibility of forming an organisation that would restore the Parliament had been present to his mind through all the vicissitudes of his career.

Did O'Connell believe in the possibility of obtaining

Repeal by agitation? To answer this question, as Gustave de Beaumont observes, but a little knowledge of human nature is required. We all know that the tendency of our minds is to underrate the difficulties of attaining any object of ambition in proportion to the duration and the enthusiasm of our desire. The lawyer after a few hours' study of a doubtful case will frequently become entirely identified with it, will persuade himself that its arguments are irresistibly cogent, and look forward with the utmost confidence to its triumph. How much more easily may the politician become overconfident in a cause which has been the dream of his life, and underestimate the obstacles in his path! It is impossible to read the published conversations of O'Connell without feeling that he was naturally of a most sanguine temperament. It is impossible to follow his career without perceiving that it was eminently calculated to foster such a temperament. He had entered into politics upon an untrodden path, with no precedent to guide him, with no encouragement to cheer him, with no experience to sustain him. The most illustrious of his fellow-countrymen had predicted his failure. He had seen public opinion among his co-religionists so faint as scarcely to be perceptible to the rulers. He had made it so terrible that the resolution of Wellington and the ability of Peel quailed beneath it. He had seen the society of his creation unable to secure the attendance of ten members at its meetings, and he had made it the ruler of Ireland. He had seen the Roman Catholic clergy equally submissive and powerless, and by their instrumentality he had wielded the passions of the nation. Looking back to such a triumph as that of 1829, encouraged by the sympathy and admiration of the leading nations of Europe, and idolised by the immense majority in his own, was it surprising that he should have entered with confidence and with cheerfulness upon the struggle? His first object was to convince the people that their efforts would be successful; and in convincing them he strengthened his own conviction. The occupation of his life for many years was to throw the Repeal arguments into the most fascinating and imposing light; and in doing so his own belief in his cause rose to fanaticism. It is related of him that he suggested that but one line should be graven upon his tomb— 'He died a Repealer.'

And was his hope so absolutely unreasonable? Was it impossible—was it even very improbable—that the Irish Parliament might have been restored? O'Connell perceived clearly that the tendency of affairs in Europe was towards the recognition of the principle that a nation's will is the one legitimate rule of its government. All rational men acknowledged that the Union was imposed on Ireland by corrupt means, contrary to the wish of one generation. O'Connell was prepared to show, by the protest of the vast majority of the people, that it was retained without the acquiescence of the next. He had allied himself with the parties that were rising surely and rapidly to power in England—with the democracy, whose gradual progress is effacing the most venerable landmarks of the constitution—with the Freetraders, whose approaching triumph he had hailed and exulted in from afar. He had perceived the possibility of forming a powerful party in Parliament, which would be free to co-operate with all English parties without coalescing with any, and might thus turn the balance of factions, and decide the fate of Ministries. He saw, too, that while England in a time of peace might resist the expressed will of the Irish nation, its policy would be necessarily modified in time of war; and he predicted that should there be a collision with France while the nation was organised as in '43, Repeal would be the immediate and the inevitable consequence. In a word, he believed that under a constitutional government the will of four-fifths of a nation, if peacefully, perseveringly, and energetically expressed, must sooner or later be triumphant. If a war had broken out during the agitation—if the life of O'Connell had been prolonged ten years longer—if any worthy successor had assumed his mantle—if a fearful famine had not broken the spirit of the people—who can say that the agitation would not have been successful? Such a contest, however, was too great to be compressed into the closing years of a laborious life.

But then we are met with the ready answer—the Repeal rent was the object of the Repeal agitation. For years this rent was the ceaseless subject of the ridicule of the writers of the British press, and placemen of every order declaimed in choicest periods on the iniquity of receiving money for political services. All this affected indignation appears to me, I confess singularly ridiculous. To suppose that a vast movement, extending over nearly the whole surface of Ireland, sending its agents to every county and to every parish, exercising its influence upon every election, collecting statistics, redressing wrongs, preparing petitions, and actively propagating its opinions, could be created and maintained without a regular tribute is palpably absurd. The Repeal rent was necessary for the maintenance of the organisation, and it was also the most imposing manifestation of its power. No equally efficacious means has ever been adopted of giving cohesion to a great political movement, or securing the sustained and intelligent co-operation of the people, of exhibiting beyond all question the extent and the intensity of the public feeling, and of proving its progressive character. To make O'Connell the recipient of the rent was the only means of making it thoroughly popular, and of preventing those disputes and recriminations that would have been so injurious to the cause. O'Connell was the idol of the nation. He had relinquished for its service a lucrative practice at the Bar; he had surrendered all hopes of promotion to the Bench, to which he would otherwise have undoubtedly attained, and where he might have spent his closing years in affluence and in dignified ease. His sacrifices, his position, and his genius rendered the tribute in the eyes of his supporters a fitting reward for his services, and a fitting testimonial of their affection. How faithfully it was expended his death sufficiently proved. Though he had been one of the most popular lawyers in Ireland when he practised at the Bar, and though he had inherited a considerable property from his uncle in 1825, he died broken in fortune as in spirits. Out of the princely revenue he had commanded he scarcely secured a competency for his children. He had received it from the people's love—he spent it in the people's cause.

To these considerations two answers are given. It is said that O'Connell lived in the most luxurious manner, keeping open house, and exercising the most unbounded hospitality, and that he also employed a large portion of the tribute in bringing his relations into Parliament. With reference to the first charge, it might be sufficient to say that a man whose life was spent for the most part in Herculean public labours might well be pardoned if, in the rare hours of relaxation, he employed every possible means of stimulating and invigorating a mind jaded by excess of toil. But there is a fuller answer than this. O'Connell was the leader of a great agitation. He had formed a system of government which he designed to exhibit as eclipsing the recognised government of Ireland. He was the centre of a vast movement which radiated over three provinces. For a man occupying such a position keeping up intimate relations with so many politicians and directing such various operations, great hospitality was absolutely necessary. No one ascribes the hospitality of the Prime Minister, or of any other political leader, to a spirit of self-indulgence. It is simply the necessity of their position. And with reference to the elevation of his relatives to Parliament, while there can be no doubt that it was gratifying to his feelings, it is by no means clear that it was injurious to his cause. His grand object, as a Parliamentary leader, compared with which every other became insignificant, was to inspire his party with perfect unanimity. In no conceivable way could he more fully effect that object than by bringing into Parliament men who were personally attached to himself. His followers were in general not very eminent or very high-minded men, but it would be difficult to show any instance in which, by procuring the election of a relative, he excluded a man of real ability.

The career of O'Connell, during the Repeal movement, divides itself into two distinct parts—his Parliamentary life and his agitation in Ireland. He readily perceived that to bring the Repeal question at once into Parliament would be extremely unwise. Parliament is, in the first instance, always almost unanimous in opposing any radical change. It is only when the public opinion has been thoroughly gained, when the evils of resistance are shown to be greater than those which can flow from concession, and when the question has assumed an overwhelming magnitude, that the Parliamentary tide turns. Its change is then often both sudden and complete. O'Connell, perceiving this, determined to abstain from discussing the subject in Parliament, and resisted very resolutely the taunts by which the English members endeavoured to urge him to a division; but a party in Ireland, represented by Feargus O'Connor and the 'Freeman's Journal,' argued so vehemently for a Parliamentary discussion that in 1834 he was at length compelled to yield. The result, as might have been easily anticipated, was an utter failure. Only one English member voted for Repeal, and the majority against it amounted to nearly 500. The division discouraged him greatly, and perhaps somewhat damped the ardour of the movement.

The real importance however of the Repeal movement was shown outside the walls of Parliament, and after the substitution of Sir R. Peel for Lord Melbourne as Prime Minister. Sir R. Peel, though one of the least fanatical, had been one of the most formidable adversaries of the Catholic claims. When Secretary for Ireland, his eulogies of the Orangemen and his exclusive promotion of anti-Catholics had earned for him the nickname of 'Orange Peel,' and he and O'Connell always regarded one another with intense enmity, both personal and political. He now declared that there was 'no influence, no power, no authority which the prerogative of the Crown and the existing laws gave the Government, that should not be exercised for the purpose of maintaining the Union.' The Chancellor Sugden dismissed from the magistracy O'Connell and some other conspicuous Repealers, and it was clearly understood that no one who held the obnoxious opinions had the slightest chance of obtaining any office from the Government or any recognition of his talents at the Bar. Some young lawyers of promise selected this time for joining the movement, and the people, whose confidence in their leader was boundless, accepted the defiance with joyful alacrity. Ireland was indeed now fully prepared for the contest. There was no hesitation, no eclecticism manifest in any party. The lines of demarcation were clearly drawn. Those vacillating and equivocal characters who were compared by O'Connell to the monsters in the 'Arabian Nights' with green backs and orange tails had nearly all disappeared. The organisation of the Repealers had been elaborated almost to perfection, and had attained its full dimensions. The Repeal Society consisted of three classes—the volunteers who subscribed or collected 10*l.* a year, the members who subscribed 1*l.*, the associates who subscribed 1*s.* The rents were collected by the instrumentality of the clergy. The unity of the organisation was maintained by Repeal wardens, under the direction of O'Connell, who presided over assigned districts. The exertions of the society were directed to the extension of Repeal influence at the elections, to the preparation of petitions, and to the assemblage of monster meetings.

O'Connell, after a time, devoted himself almost exclusively to the agitation in Ireland, and in 1843, the year of the monster meetings, he abstained altogether from Parliamentary duties. During this year he occupied perhaps the pinnacle of his glory. There are three great instances on record of politicians, discouraged by overwhelming majorities, seceding from Parliament. Grattan gave up his seat and became utterly powerless in the country. Fox retired from the debate, though retaining his seat, and he too became for a time little more than a cipher. O'Connell followed the example of Fox, but he drew with him the attention of Europe. In no previous portion of his career, not even when he had gained emancipation from the humbled Ministry of Wellington, did he attract greater attention or admiration. Whoever turns over the magazines or newspapers of the period will easily perceive how grandly his figure dominated in politics, how completely he had dispelled the indifference that had so long prevailed on Irish questions, how clearly his agitation stands forth as the great fact of the time.

It would be difficult, indeed, to conceive a more imposing demonstration of public opinion than was furnished by those vast assemblies which were held in every Catholic county, and attended by almost every adult male. They usually took place upon Sunday morning, in the open air, upon some hillside. At daybreak the mighty throng might be seen, broken into detached groups and kneeling on the greensward around their priests, while the incense rose from a hundred rude altars, and the solemn music of the Mass floated upon the gale, and seemed to impart a consecration to the cause. O'Connell stood upon a platform, surrounded by the ecclesiastical dignitaries and by the more distinguished of his followers. Before him that immense assembly was ranged without disorder, or tumult, or difficulty; organised with the most perfect skill, and inspired with the most unanimous enthusiasm. There is, perhaps, no more impressive spectacle than such an assembly, pervaded by such a spirit, and moving under the control of a single mind. The silence that prevailed through its whole extent during some portions of his address; the concordant cheer bursting from tens of thousands of voices; the rapid transitions of feeling as the great magician struck alternately each chord of passion, and as the power of sympathy, acting and reacting by the well-known law, intensified the prevailing feeling, were sufficient to carry away the most callous, and to influence the most prejudiced; the critic, in the contagious enthusiasm, almost forgot his art, and men of very calm and disciplined intellects experienced emotions the most stately eloquence of the senate had failed to produce.[1]

The greatest of all these meetings, perhaps the grandest display of the kind that has ever taken place, was held around the Hill of Tarah. According to very moderate computations, about a quarter of a million were assembled there to attest their sympathy with the movement. The spot was well chosen for the purpose. Tarah of the Kings, the seat of the ancient royalty of Ireland, has ever been regarded by the Irish people with

---

[1] The following is Bulwer's description of the scene:

> Once to my sight the giant thus was given,
> Walled by wide air and roofed by boundless heaven:
> Beneath his feet the human ocean lay,
> And wave on wave flowed into space away.
> Methought no clarion could have sent its sound
> E'en to the centre of the hosts around;
> And, as I thought, rose the sonorous swell,
> As from some church-tower swings the silvery bell;
> Aloft and clear from airy tide to tide
> It glided easy, as a bird may glide.
> To the last verge of that vast audience sent,
> It played with each wild passion as it went:
> Now stirred the uproar—now the murmurs stilled,
> And sobs or laughter answered as it willed.
> Then did I know what spells of infinite choice
> To rouse or lull has the sweet human voice.
> Then did I learn to seize the sudden clue
> To the grand troublous life antique—to view
> Under the rock-stand of Demosthenes
> Unstable Athens heave her noisy seas.—*St. Stephens.*

something of a superstitious awe. The vague legends that cluster around it, the poetry that has consecrated its past, and the massive relics of its ancient greatness that have been from time to time discovered, have invested it with an ineffable and a most fascinating grandeur. It was on this spot that O'Connell, standing by the stone where the Kings of Ireland were once crowned, sketched the coming glories of his country. Beneath him, like a mighty sea, extended the throng of listeners. They were so numerous that thousands were unable to catch the faintest echo of the voice they loved so well; yet all remained passive, tranquil, and decorous. In no instance did these meetings degenerate into mobs. They were assembled, and they were dispersed, without disorder or tumult; they were disgraced by no drunkenness, by no crime, by no excess. When the Government, in the State trials, applied the most searching scrutiny, they could discover nothing worse than that on one occasion the retiring crowd trampled down the stall of an old woman who sold gingerbread.

This absence of disorder was partly owing to the influence of O'Connell, and partly to that of Father Mathew. The extraordinary career of that wonderful man was at this time at its height, and Teetotalism was nearly as popular as Repeal. The two movements mutually assisted one another, and advanced together. The splendid success of Father Mathew was probably owing in a great measure to the fact that O'Connell had strung the minds of the people to a pitch of almost heroic enthusiasm; and, on the other hand, O'Connell declared that he would never have ventured to hold the monster meetings were it not that he had the Teetotallers ' for his policemen.' There was scarcely a Catholic county where these meetings were not held, and those who attended them have been reckoned by millions.

In the same year a very remarkable evidence was furnished of the extent to which the Repeal opinions were held by the intellect of the country in the creation of the 'Nation' newspaper. I know few more melancholy spectacles—no more mournful illustration of the declension of the national party in Ireland than is furnished by the contrast between the present of that paper and its past. What it is now it is needless to say. What it was when Gavan Duffey edited it—when Davis, Macarthy, and all their brilliant companions contributed to it, and when its columns maintained with unqualified zeal the cause of liberty and nationality in every land, Irishmen can never forget.

And over all this vast movement O'Connell at this time reigned supreme. There was no rival to his supremacy—there was no restriction to his authority. He played with the fierce enthusiasm he had aroused with the negligent ease of a master; he governed the complicated organisation he had created with a sagacity that never failed. He had made himself the focus of the attention of other lands, and the centre around which the rising intellect of his own revolved. He had transformed the whole social system of Ireland; almost reversed the relative positions of Protestants and Roman Catholics; remodelled by his influence the representative, the ecclesiastical, the educational institutions, and created a public opinion that surpassed the wildest dreams of his predecessors. Can we wonder at the proud exultation with which he exclaimed, 'Grattan sat by the cradle of his country, and followed her hearse: it was left for me to sound the resurrection trumpet, and to show that she was not dead, but sleeping'?

Among the popular methods of depreciating the intellect of O'Connell, one of the principal has been to represent him simply as a member of a very numerous and a very much despised class, who are known by the name of demagogues. Now, if by a demagogue is understood a man who is merely an adept in mob-oratory, whose life is spent in pandering to the passions of the populace, in following and in interpreting their follies, and in advocating the extreme opinions they delight in, it is quite true that such a character is a contemptible one, but equally true that it does not apply to O'Connell. The truth is, that the position of O'Connell, so far from being a common one, is absolutely unique in history. There have been many greater men, but there is no one with whom he compares disadvantageously, for he stands alone in his sphere. We may search in vain through the records of the past for any man who, without the effusion of a drop of blood, or the advantages of office or rank, succeeded in governing a people so absolutely and so long, and in creating so entirely the elements of his power. A king without rebellion, with his tribute, his government, and his deputies, he at once evaded the meshes of the law and restrained the passions of the people. He possessed to the highest degree the eloquence and the adroitness of a demagogue, but he possessed also all the sagacity of a statesman and not a little of the independence of a patriot. He yielded frequently to the wishes of the people and to the passions around him, but on points which he deemed important he was quite capable of resisting them. He believed the poor-laws to be erroneous in their principle and demoralising in their action, and he opposed them on the most unpopular grounds, though Dr. Doyle, the ablest and most popular of the Roman Catholic prelates, had come forward to advocate them. He rejected without hesitation the proffered alliance of the Chartists, though Englishmen of almost every other class were inveighing against him. He was extremely anxious to obtain the sympathy and support of American public opinion, but he did not hesitate mortally to offend a large section of the American people by the zeal with which he threw himself into the cause of negro emancipation, and by his fiery denunciations of slavery. He strongly censured the existing system of insecure tenancies, and anticipated very accurately the Bill which has recently passed, saying, on one occasion, that 'nothing will do but giving some kind of fixity of tenure to the occupier, and especially an absolute right of recompense for all substantial improvements.' But, although he often used very violent and very unjustifiable language towards individual landlords, he never encouraged those socialistic notions about land which since his death have been so prevalent; and he never forgave Arthur O'Connor for having, as he heard, a plan for the equal division of land.[1] He regarded strikes as one of the curses of the country, and in 1838, when they were very prevalent in Ireland, and were supported by numbers of his followers, he was among the most prominent of those who denounced them. On this occasion he seriously imperilled his influence. He was scarcely able to obtain a hearing at a meeting he attended. He was hooted through the streets of Dublin, but he never shrank from warning the people against those combinations, and he succeeded for a time in putting them down in Ireland.

---

[1] See O'Neil Daunt's 'Personal Recollections of O'Connell,' vol. i. p. 50; vol. ii. p. 232.

But the noblest instance of his moderation is furnished by his constant denunciations of rebellion. An orator who sought only for popularity in addressing so bellicose a people as the Irish would have dwelt constantly on the verge of treason, and have continually dilated upon the glories of the battle-field. O'Connell, on the other hand, uniformly warned the people against appealing to arms. He exhausted all his eloquence in contrasting the advantages of constitutional agitation with the horrors of war, and exhibited at all times, both in public and in private conversations, an almost Quaker detestation of force. Perhaps no higher tribute has ever been paid him than that of Mr. Mitchell, who declared that, next to the British Government, he regarded O'Connell as the greatest enemy of Ireland; for it was altogether owing to his eloquence and to his principles that the Irish people could not be induced to follow the revolutionary movement of 1848. He infused into them a touching faith in the power of peaceful agitation, which unhappily did not survive his defeat. He proclaimed himself the first apostle of that sect whose first doctrine was, that no political change was worth shedding a drop of blood, and that all might be attained by moral force; and he confidently looked forward to the time when the might of public opinion would prove invariably triumphant in political struggles. As one of the poets of the movement wrote:

> When the Lord created the earth and the sea,
>   The stars, and the glorious sun,
> The Godhead *spoke*, and the universe woke,
>   And the mighty work was done!
> Let a word be flung from the orator's tongue,
>   Or a drop from the fearless pen,
> And the chains accurst asunder burst
>   That fettered the minds of men.

> Oh! these are the arms with which we fight,
>   The swords in which we trust,
> Which no tyrant hand shall dare to brand,
>   Which time cannot stain or rust.
> When these we bore we triumphed before,
>   With these we'll triumph again,
> And the world shall say no power can stay
>   The voice or the fearless pen.[1]

The system of gigantic, organised agitation for political ends which he devised was a discovery in politics, and the example was speedily followed in England, and tended very powerfully to discredit the conspiracies and riots to which the unrepresented classes had long been prone. The Corn-Law League, which obtained for England the blessing of free trade, was in a great degree an imitation of the Catholic Association of O'Connell.

That the outrageous language he sometimes employed, his habitual use of the term Saxon instead of Englishman, and his frequent recurrence to the worst episodes of the past history of Ireland contributed much to separate the two nations is undoubted; but it must be added that, while his influence lasted, there was none of that malignant type of disloyalty which has since then been so common. The people were anti-English because of the Union or the Protestant ascendency, but they always retained a kind of reversionary loyalty, and looked forward, when their grievances were redressed, to a cordial union with England. It must be added, too, that O'Connell always drew a broad line of distinction between the

---

[1] Macarthy. Contrast the lines of the Young Ireland poet Davis:
> The tribune's tongue or poet's pen
>   May sow the seed in prostrate men,
> But 'tis the soldier's sword alone
>   Can reap the harvest when 'tis grown.

Sovereign and her Ministers, and there was probably no period of his agitation in which the Queen would not have been received with enthusiasm in Ireland. If the measures which he adopted were often very culpable, the great end of his politics will now be very generally admitted to have been good. His advocacy of universal suffrage, his crusade against the House of Lords, his ferocious denunciations of the upper classes of his fellow-countrymen, perhaps even his agitation for Repeal, were all means to an end—that end being the elevation of the Catholics from a pariah class into a position of equality with the Protestants. That policy has since been fully carried out. No one will now defend the old system of tithes, and few will question that the Appropriation Clause was just. The Church policy, which was thought so extravagant in 1833, has been carried out in 1869 with a severity which O'Connell never advocated, and the security of tenure which O'Connell claimed for the Irish tenant has been amply provided by the Land Bill of 1870.

Nor can O'Connell be justly regarded as the mere tool of the clergy. It is true that he first brought them into the political arena and governed by their means, but he was invariably the director of their policy. He refused emphatically to submit to be dictated to by his spiritual advisers. 'We are Roman Catholics,' he once said, 'but not servants of Rome;' and he fully echoed the words of his secretary, 'As much theology as you please from Rome, but no politics.' Though he was passionately attached to his own religion, and on most subjects very little apt to restrain his invective, it would be difficult or impossible to find a single instance in which he used offensive language against Protestants as such. Though perpetually confronted with the grossest Puritan bigotry, he exhibited himself a steady and large-minded tolerance for every form of religious belief, that raised him immeasurably above his Protestant adversaries. 'In plain truth,' he said, in language which there is every reason to believe expressed his deepest conviction, 'every religion is good — every religion is true to him who in his due caution and conscience believes it. There is but one bad religion, that of a man who professes a faith which he does not believe; but the good religion may be, and often is, corrupted by the wretched and wicked prejudices which admit a difference of opinion as a cause of hatred.' He continually laboured in the spirit of Grattan and O'Leary to allay the religious discord of his countrymen, accepted cordially every overture made to him by Protestants, advocated the cause of religious liberty in every quarter, and alone, of all prominent Roman Catholics, succeeded in making himself through his whole life the champion of the Church, and at the same time a consistent leader of the most advanced Liberal party. It is this aspect of his career which seems to have most struck Continental writers, and to have made him 'a representative man' in his Church.

The struggle against the Church of Rome in the present day is not strictly theological. Its real adversary is no longer the Protestant divine, nor are the weapons of the controversy those of dogmatic polemics. A new method and severity of historical criticism, by sapping the authority of the Church, and a series of momentous scientific discoveries, by familiarising men with anti-theological conceptions of the nature and government of the universe, are gradually loosening its hold upon the minds of men, while at the same time its power is immeasurably diminished by a great political change. The theological doctrine of the Divine right of kings was the basis of the government of Catholic Europe, but since the French Revolution this theological basis has been generally repudiated, the whole sphere of politics is fast passing beyond the empire of the Church, the government of a great part of Europe rests upon principles which she cannot approve, and the sympathies of the people are in habitual opposition to those of the priests. The great Liberal party that ramifies over nearly the whole of Europe, and advances side by side with education and social progress, is in open or disguised antagonism to the Church, and, as its triumph becomes every year more certain, the priestly power is waning rapidly in lands where the doctrines of Protestantism are unknown. It was the work of O'Connell to make the Liberal party, in Ireland at least, synonymous with the Catholic party. By drawing clearly the distinction between rebellion which the Church condemns, and agitation which it does not condemn; by advocating in Parliament the cause of every oppressed nationality; by claiming religious equality for the Dissenters as well as for his co-religionists; by allying himself with the most advanced democrats; and, above all, by making his cause essentially national, he succeeded in becoming at once the greatest Catholic and one of the greatest Liberals of his age. Three or four of the most gifted intellects of France were engaged at the same time, though with very indifferent success, in advocating this alliance, and they regarded O'Connell as their great model and representative. On this ground three of the most eloquent men on the Continent—Montalembert, Ventura, and Lacordaire—have made him the subject of the most splendid eulogy. The attempt to make the Catholic priesthood the representatives of sincere Liberalism has, as might have been expected, proved ultimately hopeless; but if O'Connell did not ally his cause permanently with Liberalism abroad, he at least succeeded in identifying it with Nationalism at home. He contrived to place the Protestant clergy in direct opposition to the sympathies of the people, to neutralise all the good effect of the Liberalism of Grattan or Curran, and thus to raise a formidable rampart around his Church. Religious doctrines with great masses of men depend very little for their acceptance on the unbiassed judgment of the intellect, and very much upon the sympathy and the esteem inspired by their teachers, and a Church which has sold the birthright will never obtain the blessing. The Irish Protestant Church, accepting the position of an English garrison in an enemy's country, supporting for the most part a policy of restriction and disqualification, and opposing the national aspirations of the people, has occupied a position very similar to that of the Papacy in Italy, while in Ireland, as in Poland and in Spain, Catholicism has derived an incalculable force from being the symbol of national feeling. It is probable, however, that this situation will gradually be modified. Recent measures of disestablishment and disendowment, by depriving the Protestant Church of all the privileges it derived from the State, have destroyed its invidious and exceptional position, and removed a chief obstacle to an Irish policy on the part of its members, while among the Catholic priests Ultramontanism is becoming more and more ascendant, and their policy is, in consequence, more and more evidently subservient to foreign dictation.

With the great qualities of O'Connell there were mingled great defects, which I have not attempted to conceal, and which are of a kind peculiarly repulsive to a refined and lofty nature. His character was essentially that of a Celtic peasant. Though he was the representative of an old family, and though he had received a good education in France, he exhibited to a singular degree the characteristic faults of an uneducated man — coarseness, scurrility, cunning, a power of passing on the slightest occasion from the extreme of flattery to the extreme of abuse, a looseness of statement which is not altogether explained by the natural exaggeration of the Celtic mind. Of the faults of taste into which he could fall, and the manner in which he could expose himself to ridicule, it is sufficient to say that in 1838 he published a letter describing himself as having, during a sleepless night, cried bitterly in bed because Lord J. Russell had refused to adopt the ballot. The dedication to the Queen of his memoir on the past atrocities of English Governments in Ireland is written in a strain of bombast that would disgrace the pen of the editor of a country newspaper; and there are many things in his other writings and in his speeches which are equally puerile. As was almost inevitable from his mode of life, his faults grew upon him with age. Perpetually speaking before crowds of uneducated men, and perpetually breathing the atmosphere of the most vulgar flattery, his intellect and character were alike lowered. It is indeed a grave, though a common error, to judge speeches addressed to an uneducated audience by the canons of a refined taste; for a great orator will always adapt his style to his audience, and will know that coarse humour, or florid imagery, or claptrap declamation may affect some classes more than all the eloquence of Demosthenes; but when all due allowance is made for this, it remains true that the language of

O'Connell lowered the tone of public opinion in Ireland, and the character of the nation in the eyes of the world. He represented, played upon, and strengthened some of the worst defects of the Irish nature, and there was very little that was either manly or dignified in his later oratory. At the same time, his violence was sometimes almost ungovernable. He often complained with justice of landlord intimidation as applied to voters, but it would be difficult or impossible to find instances of more scandalous intimidation than was practised by his followers at his instigation in 1835;[1] and the language he habitually employed towards his opponents gave a bitterness to political controversy in Ireland which it had never before attained. One of the most hopeful circumstances in the present condition of the country is that the generation is fast passing away which rose to manhood during his agitation. Few generations of Irishmen have exhibited so little real genius. None has been so profoundly divided by sectarian and party hatred.

The result of his career was in another way profoundly injurious to the country. The main object of the legislative Union had been to withdraw the Government of Ireland from the hands of the Irish gentry, and one of its most important results was to diminish their influence as the political leaders of the people. By a singular fatality, the great advocate of Repeal continued this policy, and thus did more than anyone else to make the Union a necessity. From the beginning of his career, when he crushed the influence of the leading Irish lay Catholics on the question of the Veto, to the end of his struggle for Repeal, he was continually employed in breaking or weakening

---

[1] See the 'Annual Register,' 1835, p. 15.

the landed classes, in dispelling the feudal reverence of the people, and in making the priests their political leaders. In the case of individual landlords, indeed, he often showed himself anxious to conciliate, and even fulsome in his adulation,[1] but he destroyed the sympathy between the people and their natural leaders; and he threw the former into the hands of men who have subordinated all national to ecclesiastical considerations, or into the hands of reckless, ignorant, and dishonest adventurers. If the people and the possessors of property in Ireland were now cordially united they could obtain any measure of self-government they desired, and the Ultramontane policy dictated by the priesthood, and the wild socialistic follies of Fenianism, are the chief obstacles to its attainment. No truths can be more obvious than that a cordial union between Irishmen of all creeds is the first condition of political progress in Ireland, and that a demand for any measure of self-government must rest upon the doctrine that the public opinion of a country should determine the form of its government; but one section of the popular leaders in Ireland are now straining every nerve to break down the system of united education, which is the best hope for the future of their country, and to incline the foreign policy of the empire to the side of everything that is anti-Liberal on the Continent, while another section are advocating doctrines subversive of those fundamental rights of property which it is a main end of all government to secure, and a policy of rebellion which, if it could be realised, could be realised only at the expense of a massacre of their

---

[1] The reader may find some curious instances of this in Lord Cloncurry's 'Personal Recollections.' It was a very characteristic saying of O'Connell that 'you may catch more flies with a spoonful of honey than with a hogshead of vinegar.'

fellow-countrymen. If at the present moment the antagonism of classes and creeds is stronger in Ireland than in any other country in Europe—if there is no part of the empire in which genuine, modest, and manly talent is so little appreciated by constituencies, and in which the demagogue and the adventurer can find so favourable a field—this is to be mainly attributed to the policy of Pitt, and to the agitation of O'Connell. By grave faults on both sides the natural ties that united classes have been broken, and until they are in some degree formed anew there is never likely to be a consistent and successful national policy in Ireland.

I have dwelt at considerable length upon the faults and merits of O'Connell, for the position I would venture to assign him is much higher than that which is usually conceded him in England, and there are few men who are estimated more differently in England and on the Continent. It is impossible to judge his position without taking into account the place he occupied with reference to the progressive party in his Church, and the depreciatory tone adopted by his many enemies has naturally made a deep impression on the public mind. There is also a constant tendency —especially among intellectual people—to underrate those whose genius is employed chiefly in action, especially when the lower orders are subjects of that action.

If I were asked to point out a personage in history who in intellectual and moral temperament bore a striking resemblance to O'Connell, I should select one who differed from him in principles as widely as any that could be named, and who has played a far greater and far nobler part in the affairs of men—I mean Martin Luther. There is something in the very

appearance of these men exhibiting the same nature—
a nature of indomitable strength, genial rather than
refined, massive and precious, but somewhat coarse-
grained. In each, character and intellect so happily
harmonised that it were hard to say how much their
success was due to force of will, and how much to
force of mind. In each was the same instinctive tact
in governing great masses of men, the same calculated
audacity, the same intuitive perception of opportunities,
the same art in inspiring and in retaining confidence.
Each displayed an eloquence of the most popular
character, nervous, pointed, but incorrect; thrilling
and fascinating, by the glow of feeling that pervaded
it; repelling and irritating, by the coarseness, the
vituperation, the vulgarity into which it degenerated.
Each was associated with men of purer intellectuality
and more heroic enthusiasm, yet each, if measured by
his achievements, towers above all his associates. Neither
can be judged fairly by a microscopic and a detailed
criticism. It is easy to detect acts that cannot be
justified, language that can scarcely be palliated, in-
consistencies that it is difficult to explain. But, though
their opponents will never be at a loss for subject-
matter for their attacks, though their admirers will
always find much that they must deeply deplore, and
though the sentimentalist will turn with disgust from
men in whose temperaments the grosser elements so
largely mingled, yet the stamp of true genius is upon
both, and the aureole that marks those who have
laboured faithfully for mankind will ever circle their
memories. The magnitude and the unity of their
lives become only visible when distance has enabled
the eye to discover their full proportions, and when
experience has shown how miserable were the efforts of
their successors to wield their sceptres. Nay, in the

very inequalities of their tempers there is much to attract sympathy. Luther, hurling his unmeasured invective against some royal opponent, and then pouring out a strain of the gentlest tenderness over his child—O'Connell, listening with calm complacency to the crowd of orators who 'were advertising' him by their denunciations, yet galled to the quick by the sarcasm of an old friend—present a resemblance as pleasing as it is striking. Both were men of powerful intellects and of warm hearts, and both, with great though unequal faults, laboured with a firm faith to realise objects which they believed to be good.[1]

The Government was extremely alarmed at the success of the monster meetings, and they at length determined by a bold measure to crush the agitation. A meeting had been advertised for Sunday, October 3, 1843, to be held at Clontarf. It would have been probably one of the very greatest of the series, for Clontarf is in the immediate vicinity of Dublin. The meeting had been announced about a fortnight before. The Government took no notice of it till the afternoon of the 2nd, when the roads were thronged with the excited populace, who had come from a distance to attend it, and a proclamation was then issued forbidding it. It is said that the cannon of 'the Pigeon-house' were actually turned upon Clontarf. The natural consequence of this proceeding of the Government of Sir R. Peel would have been a breach of the peace and a massacre more sanguinary than that of Manchester, and this would almost certainly have taken place but

---

[1] 'Oh for a great man,' said Coleridge, 'but one really great man, who could feel the weight and the power of a principle, and unflinchingly put it into act! See how triumphant in debate and action O'Connell is. Why? Because he asserts a broad principle, and acts up to, rests all his body on it, and has faith in it.'—*Table Talk*.

for the extraordinary promptitude of O'Connell. He at once despatched messengers in all directions to apprise the people, and by exerting all his wonderful influence induced them peaceably to disperse.

The Government prosecution followed close on the proclamation. It was a charge of conspiracy, or, in other words, of the employment of seditious language, against O'Connell, his son, and five of his principal followers. The trial was extremely protracted; but its monotony was relieved by much brilliant oratory, by a great deal of very curious cross-examination, and by an amusing episode occasioned by the Attorney-General, who sent a challenge to one of the opposing counsel, which that gentleman submitted to the Bench. The two most eloquent speeches delivered were beyond all question those of Sheil and Mr. Whiteside. A great number of charges have been brought against this trial which have elicited much controversy. It is sufficient to state the facts that are admitted. An error, which at least one Irish judge believed not to have been unintentional, was made in the panel of the jury, and by this error more than twenty Catholics were excluded from the juror list. Of the Catholics whose names were called all were objected to by the Government prosecutor, and accordingly there was not a single Roman Catholic on the jury which tried the greatest Catholic of his age in the metropolis of an essentially Catholic country, and at a time when sectarian animosity was at its height. After a charge from the Chief Justice which Macaulay afterwards compared to the displays of judicial partisanship in the State trials of Charles II., O'Connell was found guilty, and condemned to two years' imprisonment, together with a fine—a sentence against which he appealed to the Lords.

Some months elapsed before the appeal could be heard, and during the earlier part of that time O'Connell was in great, though, as it proved, needless alarm, lest the people should have broken into open rebellion. He despatched from prison the most emphatic addresses, exhorting them to tranquillity, and he soon found that they were quite willing to respond to his appeal. Their reception of the Government prosecution was very striking. They remained perfectly tranquil; but the rent, which in the fourteen weeks before the trial had been 6,679*l.*, rose in the fourteen weeks that followed it to 25,712*l.* In the first week it was nearly 2,600*l.*

At the beginning of the trial, Mr. Smith O'Brien gave for the first time his formal adhesion to the movement, and, during the imprisonment of O'Connell, the leadership of the party devolved upon him. Though very deficient, both in oratorical abilities and in judgment, he obtained great weight with the people from the charm that ever hangs around a chivalrous and polished gentleman, and from the transparent purity of a patriotism on which suspicion has never rested; and he was also a skilful and a ready writer. Of the wisdom he displayed in one unhappy episode of his career there are not likely to be two opinions, but it should not be forgotten that it was the ceaseless labour of his life to inculcate the importance of self-reliance, to dissociate the national cause from the claptrap and the bombast by which it was so frequently disfigured, and to teach the people that Liberal politics are only truly adopted when they are applied without respect of persons and without fear of consequences. It was thus that he laboured during the lifetime of O'Connell to check the place-hunting and the boasting that disgraced the Repeal cause, and that near the close of his life he calmly and fearlessly risked all the popularity which years of suffering had gained him, by opposing those who sought to identify Irish Liberalism with Italian despotism, and to draw down upon their country the horrors of a French invasion. Few politicians have sacrificed more to what they believed to be right, and the invariable integrity of his motives has more than redeemed the errors of his judgment.

The appeal to the House of Lords was heard in September 1844. On occasions of this kind, when the House sits to review the decisions of the law courts, it is customary to leave the matter entirely in the hands of the Law Lords, and the permanent maintenance of the judicial authority of the House obviously depends upon the observance of this custom; but there have been instances in which Lay Lords have taken part in the decision.[1] O'Connell had always been the bitter enemy of the House of Lords. He had inveighed against it in the grossest terms. He had given many of its members cause for the deepest personal animosity. When the appeal was to be heard, a number of Lay Lords came down to the House to vote against him. The five Law Lords, who were present, first delivered their opinions— two of them confirming the sentence of the Irish court, three of them condemning it. Lord Denman, in the course of his judgment, stigmatised the proceedings in Ireland in the strongest language. When the Law Lords had delivered their judgment, Lord Wharncliffe rose and appealed to the other members of the House not to permit their personal or political feelings to influence a judicial sentence. The appeal struck the right chord. The high and honourable feeling that has almost always characterised the House of Lords reasserted its sway. Every Lay

[1] E.g. in the famous Douglass case in 1769.

Lord left the House, and their bitterest living enemy was freed by their forbearance.

The news of the reversal of the sentence was received in Ireland with a burst of the most enthusiastic acclamations — bonfires blazed over the country — O'Connell passed through the streets of Dublin in a triumphal procession. A perfect delirium of excitement prevailed among his followers; yet, notwithstanding these ebullitions, the spell of his power was in a great measure broken. It was said that the months of imprisonment he had undergone had shattered his health and impaired his energies. For the first time for many years, serious dissensions arose among his followers. The Young Ireland party exercised considerable influence, and appeared to exercise far more from the great talent it displayed. The 'Nation' newspaper espoused its cause. It possessed also one very brilliant orator, Thomas Francis Meagher, a young man whose eloquence was beyond comparison superior to that of any other rising speaker in the country, and who, had he been placed in circumstances favourable to the development of his talent, might perhaps have at length taken his place among the great orators of Ireland. The Young Irelanders, like the leaders in the Rebellion of 1798, were chiefly Protestants — very young, and very enthusiastic men. They differed in the first place from O'Connell on the question whether Repealers should accept offices or promotion from the Government. They argued that those who had done so had invariably abandoned the cause — that a place-hunting spirit had crept into the society — that the sordid and corrupt element it produced was actually very great, and the discredit and suspicion it attracted much greater. On the other hand, O'Connell maintained that some concessions were necessary to the maintenance of the movement in its full extent—that the possession of place was the possession of power, and that it would be peculiarly inconsistent in Repealers to refuse it, because one of their great grievances had always been that the Government uniformly confined its bounties to their opponents. But the great characteristic of the Young Ireland party was its advocacy of rebellion. It was far more independent of the priests than O'Connell, and was little swayed by theological censures, and its sympathies were more with 1798 than with 1782. It was thus (to take but one instance from many) that Meagher declared in one of his speeches, 'There are but two plans for our consideration—the one within the law, the other without the law. Let us take the latter. I will then ask you, Is an insurrection practicable? Prove to me that it is, and I for one will vote for it this very night. You know well, my friends, that I am not one of those tame moralists who say that liberty is not worth a drop of blood. Men who subscribe to such a maxim are fit for out-of-door relief, and for nothing better. Against this miserable maxim the noblest virtue that has saved and sanctified humanity appears in judgment. From the blue waters of the Bay of Salamis—from the valley over which the sun stood still and lit the Israelites to victory—from the cathedral in which the sword of Poland has been sheathed in the shroud of Kosciusko—from the convent of St. Isidore, where the fiery hand that rent the ensign of St. George upon the plains of Ulster has crumbled into dust—from the sands of the desert, where the wild genius of the Algerine so long has scared the eagle of the Pyrenees—from the ducal palace in this kingdom, where the memory of the gallant Geraldine enhances more than royal favour the nobility of his race—from the solitary grave within this mute city which a dying request has left without an epitaph—oh! from every spot where heroism has had a sacrifice or a triumph, a voice breaks in upon the cringing crowd that cherishes this maxim, crying out, Away with it! away with it!'

It will be remembered that the maxim thus denounced was one which O'Connell lost no opportunity of extolling.

The influence of the Young Irelanders was more apparent than real, for when the appeal to arms was actually made it proved absolutely impotent against the principles with which O'Connell had leavened the people. This dissension, however, greatly injured the Repeal cause. One of those reactions of despondency to which all popular movements are liable began, and the disputes about the Federal scheme in 1844 still further weakened the popular enthusiasm.

These disputes preyed greatly on O'Connell's mind, and the period that followed his release presents a confused and chaotic picture, very unlike that of former years. His health suddenly gave way. Ceaseless labour and excessive care had broken a constitution that was naturally of Herculean strength. His voice, which had once pealed with such thrilling power over assembled thousands, sank into an almost inaudible whisper. His hopes, which had once been so buoyant that they rose above all obstacles, began now to fail. Famine came with fearful rapidity upon the land, and O'Connell foresaw the evil, while he could not avert it. The chill of death was upon him—the certainty of failure wrung his soul with an agony the more bitter because of the sanguine hope that had preceded it. An unutterable, unmitigated gloom sank upon his mind, and withered and destroyed his energies. Weak and prostrate in health and hope, he attended for the last time that Legislature which he had so triumphantly entered. In a speech of simple and touching eloquence, entirely free from every tinge of his ancient violence, he showed the fearful magnitude of the calamity impending over the country, suggested his remedies, and with a solemn and heartfelt pathos implored the generous aid of Parliament. But his voice was so faint that but few could catch his words. The fearful change impressed all who saw him.[1] Old rancour and party spirit were forgotten at the spectacle of so great a sorrow. He was listened to with an almost reverential silence, and followed by many evidences of pity and of respect. Statesmen of all parties testified their sympathy by their enquiries. The Queen, with a graceful kindness that should never be forgotten, sent to ask after the dying agitator. Another visit he received in those last dark days which he must have valued still more—three of the Oxford converts to Rome came to assure him that it was his career that had first directed their attention to the theology of his Church.

Religion was indeed now the only solace of his mind. In his youth he had been dissipated and immoral; but a change had passed over him, it is said, about the time of his duel with d'Esterre, and his attachment to his religion was sincere and fervent. His physicians having ordered him abroad, he resolved to draw his last breath near the tombs of the Apostles in that great city which is the metropolis of his Church. The deep melancholy which the consciousness of the famine impending over his country produced attended him on that dreary journey. 'He seemed,' said one who visited him in France, 'to be a continued prey to sad

[1] See the very touching description in Disraeli's 'Life of Lord J. Bentinck.'

reflections. His face had grown thin, and his look proclaimed an inexpressible sadness: the head hung upon the breast, and the entire person of the invalid, formerly so imposing, was greatly weighed down.' His strength failed him when he arrived at Genoa, and in that city he expired on May 15, 1847.

He bequeathed his body to Ireland and his heart to the Eternal City. The former rests in the cemetery of Glasnevin, in the vicinity of Dublin; the latter near the tomb of Lascaris, in the church of St. Agatha, at Rome.

There is something almost awful in so dark a close of so brilliant a career. The more I dwell upon the subject, the more I am convinced of the splendour and originality of the genius and of the sterling character of the patriotism of O'Connell, in spite of the calumnies that surround his memory, and the many and grievous faults that obscured his life. But when to the great services he rendered to his country we oppose the sectarian and class warfare that resulted from his policy, the fearful elements of discord he evoked, and which he alone could in some degree control, it may be questioned whether his life was a blessing or a curse to Ireland.

AUGUST 1878.

GENERAL LISTS OF NEW WORKS

PUBLISHED BY

Messrs. LONGMANS, GREEN & CO.

PATERNOSTER ROW, LONDON.

HISTORY, POLITICS, HISTORICAL MEMOIRS &c.

Armitage's Childhood of the English Nation. Fcp. 8vo. 2s. 6d.
Arnold's Lectures on Modern History. 8vo. 7s. 6d.
Buckle's History of Civilisation. 3 vols. crown 8vo. 24s.
Chesney's Indian Polity. 8vo. 21s.
— Waterloo Lectures. 8vo. 10s. 6d.
Digby's Famine Campaign in India. 2 vols. 8vo. 32s.

Epochs of Ancient History :—
    Beesly's Gracchi, Marius, and Sulla, 2s. 6d.
    Capes's Age of the Antonines, 2s. 6d.
    — Early Roman Empire, 2s. 6d.
    Cox's Athenian Empire, 2s. 6d.
    — Greeks and Persians, 2s. 6d.
    Curteis's Rise of the Macedonian Empire, 2s. 6d.
    Ihne's Rome to its Capture by the Gauls, 2s. 6d.
    Merivale's Roman Triumvirates, 2s. 6d.
    Sankey's Spartan and Theban Supremacies, 2s. 6d.

Epochs of English History :—
    Browning's Modern England, 1820–1875, 9d.
    Cordery's Struggle against Absolute Monarchy, 1603–1688, 9d.
    Creighton's (Mrs.) England a Continental Power, 1066–1216, 9d.
    Creighton's (Rev. M.) Tudors and the Reformation, 1485–1603, 9d.
    Rowley's Rise of the People, 1215–1485, 9d.
    Rowley's Settlement of the Constitution, 1688–1778, 9d.
    Tancock's England during the American & European Wars, 1778–1820, 9d.
    York-Powell's Early England to the Conquest, 1s.

Epochs of Modern History :—
    Church's Beginning of the Middle Ages, 2s. 6d.
    Cox's Crusades, 2s. 6d.
    Creighton's Age of Elizabeth, 2s. 6d.
    Gairdner's Houses of Lancaster and York, 2s. 6d.
    Gardiner's Puritan Revolution, 2s. 6d.
    — Thirty Years' War, 2s. 6d.
    Hale's Fall of the Stuarts, 2s. 6d.
    Johnson's Normans in Europe, 2s. 6d.

London, LONGMANS & CO.

Epochs of Modern History—*continued.*
      Ludlow's War of American Independence, 2*s.* 6*d.*
      Morris's Age of Queen Anne, 2*s.* 6*d.*
      Seebohm's Protestant Revolution, 2*s.* 6*d.*
      Stubbs's Early Plantagenets, 2*s.* 6*d.*
      Warburton's Edward III., 2*s.* 6*d.*
Froude's English in Ireland in the 18th Century. 3 vols. 8vo. 48*s.*
— History of England. 12 vols. 8vo. £8. 18*s.* 12 vols. crown 8vo. 72*s.*
Gairdner's Richard III. and Perkin Warbeck. Crown 8vo. 10*s.* 6*d.*
Gardiner's England under Buckingham and Charles I., 1624–1628. 2 vols. 8vo. 24*s.*
— Personal Government of Charles I., 1628–1637. 2 vols. 8vo. 24*s.*
Greville's Journal of the Reigns of George IV. & William IV. 3 vols. 8vo. 36*s.*
Howorth's History of the Mongols. VOL. I. Royal 8vo. 28*s.*
Ihne's History of Rome. 3 vols. 8vo. 45*s.*
Lecky's History of England. Vols. I. & II., 1700–1760. 8vo. 36*s.*
— — — European Morals. 2 vols. crown 8vo. 16*s.*
— Spirit of Rationalism in Europe. 2 vols. crown 8vo. 16*s.*
Lewes's History of Philosophy. 2 vols. 8vo. 32*s.*
Longman's Lectures on the History of England. 8vo. 15*s.*
— Life and Times of Edward III. 2 vols. 8vo. 28*s.*
Macaulay's Complete Works. 8 vols. 8vo. £5. 5*s.*
— History of England :—
    Student's Edition. 2 vols. cr. 8vo. 12*s.* | Cabinet Edition. 8 vols. post 8vo. 48*s.*
    People's Edition. 4 vols. cr. 8vo. 16*s.* | Library Edition. 5 vols. 8vo. £4.
Macaulay's Critical and Historical Essays. Cheap Edition. Crown 8vo. 3*s.* 6*d.*
    Cabinet Edition. 4 vols. post 8vo. 24*s.* | Library Edition. 3 vols. 8vo. 36*s.*
    People's Edition. 2 vols. cr. 8vo. 8*s.* | Student's Edition. 1 vol. cr. 8vo. 6*s.*
May's Constitutional History of England. 3 vols. crown 8vo. 18*s.*
— Democracy in Europe. 2 vols. 8vo. 32*s.*
Merivale's Fall of the Roman Republic. 12mo. 7*s.* 6*d.*
— General History of Rome, B.C. 753—A.D. 476. Crown 8vo. 7*s.* 6*d.*
— History of the Romans under the Empire. 8 vols. post 8vo. 48*s.*
Prothero's Life of Simon de Montfort. Crown 8vo. 9*s.*
Rawlinson's Seventh Great Oriental Monarchy—The Sassanians. 8vo. 28*s.*
— Sixth Oriental Monarchy—Parthia. 8vo. 16*s.*
Seebohm's Oxford Reformers—Colet, Erasmus, & More. 8vo. 14*s.*
Sewell's Popular History of France. Crown 8vo. 7*s.* 6*d.*
Short's History of the Church of England. Crown 8vo. 7*s.* 6*d.*
Smith's Carthage and the Carthaginians. Crown 8vo. 10*s.* 6*d.*
Taylor's Manual of the History of India. Crown 8vo. 7*s.* 6*d.*
Todd's Parliamentary Government in England. 2 vols. 8vo. 37*s.*
Trench's Realities of Irish Life. Crown 8vo. 2*s.* 6*d.*
Walpole's History of England. Vols. I. & II. 8vo. (*In October.*)

BIOGRAPHICAL WORKS.

Burke's Vicissitudes of Families. 2 vols. crown 8vo. 21*s.*
Cates's Dictionary of General Biography. Medium 8vo. 25*s.*
Gleig's Life of the Duke of Wellington. Crown 8vo. 6*s.*

London, LONGMANS & CO.

Jerrold's Life of Napoleon III. Vols. I. to III. 8vo. price 18s. each.
Jones's Life of Admiral Frobisher. Crown 8vo. 6s.
Lecky's Leaders of Public Opinion in Ireland. Crown 8vo. 7s. 6d.
Life (The) of Sir William Fairbairn. Crown 8vo. 2s. 6d.
Life (The) of Bishop Frampton. Crown 8vo. 10s. 6d.
Life (The) and Letters of Lord Macaulay. By his Nephew, G. Otto Trevelyan, M.P. Cabinet Edition, 2 vols. post 8vo. 12s. Library Edition, 2 vols. 8vo. 36s.
Maguire's Pope Pius IX. Crown 8vo. 6s. Post 8vo. 2s. 6d.
Marshman's Memoirs of Havelock. Crown 8vo. 3s. 6d.
Memorials of Charlotte Williams-Wynn. Crown 8vo. 10s. 6d.
Mendelssohn's Letters. Translated by Lady Wallace. 2 vols. cr. 8vo. 5s. each.
Mill's (John Stuart) Autobiography. 8vo. 7s. 6d.
Newman's Apologia pro Vita Sua. Crown 8vo. 6s.
Nohl's Life of Mozart. Translated by Lady Wallace. 2 vols. crown 8vo. 21s.
Pattison's Life of Casaubon. 8vo. 18s.
Spedding's Letters and Life of Francis Bacon. 7 vols. 8vo. £4. 4s.
Stephen's Essays in Ecclesiastical Biography. Crown 8vo. 7s. 6d.
Stigand's Life, Works, &c. of Heinrich Heine. 2 vols. 8vo. 28s.
Zimmern's Life and Philosophy of Schopenhauer. Post 8vo. 7s. 6d.
— — — Works of Lessing. Crown 8vo. 10s. 6d.

CRITICISM, PHILOSOPHY, POLITY &c.

Amos's View of the Science of Jurisprudence. 8vo. 18s.
— Primer of the English Constitution. Crown 8vo. 6s.
Arnold's Manual of English Literature. Crown 8vo. 7s. 6d.
Bacon's Essays, with Annotations by Whately. 8vo. 10s. 6d.
— Works, edited by Spedding. 7 vols. 8vo. 73s. 6d.
Bain's Logic, Deductive and Inductive. Crown 8vo. 10s. 6d.
      PART I. Deduction, 4s. | PART II. Induction, 6s. 6d.
Blackley's German and English Dictionary. Post 8vo. 7s. 6d.
Bolland & Lang's Aristotle's Politics. Crown 8vo. 7s. 6d.
Bullinger's Lexicon and Concordance to the New Testament. Medium 8vo. 30s.
Comte's System of Positive Polity, or Treatise upon Sociology, translated:—
    VOL. I. General View of Positivism and its Introductory Principles. 8vo. 21s.
    VOL. II. Social Statics, or the Abstract Laws of Human Order. 14s.
    VOL. III. Social Dynamics, or General Laws of Human Progress. 21s.
    VOL. IV. Theory of the Future of Man; with Early Essays. 24s.
Congreve's Politics of Aristotle; Greek Text, English Notes. 8vo. 18s.
Contanseau's Practical French & English Dictionary. Post 8vo. 7s. 6d.
— Pocket French and English Dictionary. Square 18mo. 3s. 6d.
Dowell's Sketch of Taxes in England. VOL. I. to 1642. 8vo. 10s. 6d.
Farrar's Language and Languages. Crown 8vo. 6s.
Grant's Ethics of Aristotle, Greek Text, English Notes. 2 vols. 8vo. 32s.
Hodgson's Philosophy of Reflection. 2 vols. 8vo. 21s.
Kalisch's Historical and Critical Commentary on the Old Testament; with a New Translation. Vol. I. *Genesis*, 8vo. 18s. or adapted for the General Reader, 12s. Vol. II. *Exodus*, 15s. or adapted for the General Reader, 12s. Vol. III. *Leviticus*, Part I. 15s. or adapted for the General Reader, 8s. Vol. IV. *Leviticus*, Part II. 15s. or adapted for the General Reader, 8s.

London, LONGMANS & CO.

Latham's Handbook of the English Language. Crown 8vo. 6*s.*
— English Dictionary. 1 vol. medium 8vo. 24*s.* 4 vols. 4to. £7.
Lewis on Authority in Matters of Opinion. 8vo. 14*s.*
Liddell & Scott's Greek-English Lexicon. Crown 4to. 36*s.*
— — — Abridged Greek-English Lexicon. Square 12mo. 7*s.* 6*d.*
Longman's Pocket German and English Dictionary. 18mo. 5*s.*
Macaulay's Speeches corrected by Himself. Crown 8vo. 3*s.* 6*d.*
Macleod's Economical Philosophy. Vol. I. 8vo. 15*s.* Vol. II. Part I. 12*s.*
Mill on Representative Government. Crown 8vo. 2*s.*
— — Liberty. Post 8vo. 7*s.* 6*d.* Crown 8vo. 1*s.* 4*d.*
Mill's Dissertations and Discussions. 4 vols. 8vo. 46*s.* 6*d.*
— Essays on Unsettled Questions of Political Economy. 8vo. 6*s.* 6*d.*
— Examination of Hamilton's Philosophy. 8vo. 16*s.*
— Logic, Ratiocinative and Inductive. 2 vols. 8vo. 25*s.*
— Phenomena of the Human Mind. 2 vols. 8vo. 28*s.*
— Principles of Political Economy. 2 vols. 8vo. 30*s.* 1 vol. cr. 8vo. 5*s.*
— Utilitarianism. 8vo. 5*s.*
Müller's (Max) Lectures on the Science of Language. 2 vols. crown 8vo. 16*s.*
Rich's Dictionary of Roman and Greek Antiquities. Crown 8vo. 7*s.* 6*d.*
Roget's Thesaurus of English Words and Phrases. Crown 8vo. 10*s.* 6*d.*
Sandars's Institutes of Justinian, with English Notes. 8vo. 18*s.*
Swinbourne's Picture Logic. Post 8vo. 5*s.*
Thomson's Outline of Necessary Laws of Thought. Crown 8vo. 6*s.*
Tocqueville's Democracy in America, translated by Reeve. 2 vols. crown 8vo. 16*s.*
Twiss's Law of Nations, 8vo. in Time of Peace, 12*s.* in Time of War, 21*s.*
Whately's Elements of Logic. 8vo. 10*s.* 6*d.* Crown 8vo. 4*s.* 6*d.*
— — Rhetoric. 8vo. 10*s.* 6*d.* Crown 8vo. 4*s.* 6*d.*
— English Synonymes. Fcp. 8vo. 3*s.*
White & Riddle's Large Latin-English Dictionary. 4to. 28*s.*
White's College Latin-English Dictionary. Medium 8vo. 15*s.*
— Junior Student's Complete Latin-English and English-Latin Dictionary. Square 12mo. 12*s.*
Separately { The English-Latin Dictionary, 5*s.* 6*d.*
The Latin-English Dictionary, 7*s.* 6*d.*
White's Middle-Class Latin-English Dictionary. Fcp. 8vo. 3*s.*
Williams's Nicomachean Ethics of Aristotle translated. Crown 8vo. 7*s.* 6*d*
Yonge's Abridged English-Greek Lexicon. Square 12mo. 8*s.* 6*d.*
— Large English-Greek Lexicon. 4to. 21*s.*
Zeller's Socrates and the Socratic Schools. Crown 8vo. 10*s.* 6*d.*
— Stoics, Epicureans, and Sceptics. Crown 8vo. 14*s.*
— Plato and the Older Academy. Crown 8vo. 18*s.*

MISCELLANEOUS WORKS & POPULAR METAPHYSICS.

Arnold's (Dr. Thomas) Miscellaneous Works. 8vo. 7*s.* 6*d.*
Bain's Emotions and the Will. 8vo. 15*s.*
— Mental and Moral Science. Crown 8vo. 10*s.* 6*d.* Or separately: Part I. Mental Science, 6*s.* 6*d.* Part II. Moral Science, 4*s.* 6*d.*
— Senses and the Intellect. 8vo. 15*s.*

London, LONGMANS & CO.

General Lists of New Works. 5

Buckle's Miscellaneous and Posthumous Works. 3 vols. 8vo. 52s. 6d.
Carpenter on Mesmerism, Spiritualism, &c. Crown 8vo. 5s.
Conington's Miscellaneous Writings. 2 vols. 8vo. 28s.
Froude's Short Studies on Great Subjects. 3 vols. crown 8vo. 18s.
German Home Life; reprinted from *Fraser's Magazine*. Crown 8vo. 6s.
Hume's Essays, edited by Greene & Grose. 2 vols. 8vo. 28s.
— Treatise of Human Nature, edited by Green & Grose. 2 vols. 8vo. 28s.
Macaulay's Miscellaneous Writings. 2 vols. 8vo. 21s. 1 vol. crown 8vo. 4s. 6d.
— Writings and Speeches. Crown 8vo. 6s.
Mill's Analysis of the Phenomena of the Human Mind. 2 vols. 8vo. 28s.
— Subjection of Women. Crown 8vo. 6s.
Müller's (Max) Chips from a German Workshop. 4 vols. 8vo. 58s.
Mullinger's Schools of Charles the Great. 8vo. 7s. 6d.
Owen's Evenings with the Skeptics. Crown 8vo. [*Just ready.*
Rogers's Defence of the Eclipse of Faith Fcp. 8vo. 3s. 6d.
— Eclipse of Faith. Fcp. 8vo. 5s.
Selections from the Writings of Lord Macaulay. Crown 8vo. 6s.
Sydney Smith's Miscellaneous Works. Crown 8vo. 6s.
The Essays and Contributions of A. K. H. B. Crown 8vo.
    Autumn Holidays of a Country Parson. 3s. 6d.
    Changed Aspects of Unchanged Truths. 3s. 6d.
    Common-place Philosopher in Town and Country. 3s. 6d.
    Counsel and Comfort spoken from a City Pulpit. 3s. 6d.
    Critical Essays of a Country Parson. 3s. 6d.
    Graver Thoughts of a Country Parson. Three Series, 3s. 6d. each.
    Landscapes, Churches, and Moralities. 3s. 6d.
    Leisure Hours in Town. 3s. 6d.
    Lessons of Middle Age. 3s. 6d.
    Present-day Thoughts. 3s. 6d.
    Recreations of a Country Parson. Three Series, 3s. 6d. each.
    Seaside Musings on Sundays and Week-Days. 3s. 6d.
    Sunday Afternoons in the Parish Church of a University City. 3s. 6d.
Wit and Wisdom of the Rev. Sydney Smith. 16mo. 3s. 6d.

ASTRONOMY, METEOROLOGY, POPULAR GEOGRAPHY &c.

Dove's Law of Storms, translated by Scott. 8vo. 10s. 6d.
Hartley's Air and its Relations to Life. Small 8vo. 6s.
Herschel's Outlines of Astronomy. Square crown 8vo. 12s.
Keith Johnston's Dictionary of Geography, or Gazetter. 8vo. 42s.
Nelson's Work on the Moon. Medium 8vo. 31s. 6d.
Proctor's Essays on Astronomy. 8vo. 12s.
   — Larger Star Atlas. Folio, 15s. or Maps only, 12s. 6d.
   — Moon. Crown 8vo. 10s. 6d.
   — New Star Atlas. Crown 8vo. 5s.
   — Orbs Around Us. Crown 8vo. 7s. 6d.
   — Other Worlds than Ours. Crown 8vo. 10s. 6d.
   — Saturn and its System. 8vo. 14s.
   — Sun. Crown 8vo. 14s.
   — Transits of Venus, Past and Coming. Crown 8vo. 8s. 6d.
   — Treatise on the Cycloid and Cycloidal Curves. Crown 8vo. 10s. 6d.

London, LONGMANS & CO.

Proctor's Universe of Stars. 8vo. 10s. 6d.
Schellen's Spectrum Analysis. 8vo. 28s.
Smith's Air and Rain. 8vo. 24s.
The Public Schools Atlas of Ancient Geography. Imperial 8vo. 7s. 6d.
— — — Atlas of Modern Geography. Imperial 8vo. 5s.
Webb's Celestial Objects for Common Telescopes. New Edition in the press.

NATURAL HISTORY & POPULAR SCIENCE.

Arnott's Elements of Physics or Natural Philosophy. Crown 8vo. 12s. 6d.
Brande's Dictionary of Science, Literature, and Art. 3 vols. medium 8vo. 63s.
Decaisne and Le Maout's General System of Botany. Imperial 8vo. 31s. 6d.
Evans's Ancient Stone Implements of Great Britain. 8vo. 28s.
Ganot's Elementary Treatise on Physics, by Atkinson. Large crown 8vo. 15s.
— Natural Philosophy, by Atkinson. Crown 8vo. 7s. 6d.
Gore's Art of Scientific Discovery. Crown 8vo. [*In the press.*
Grove's Correlation of Physical Forces. 8vo. 15s.
Hartwig's Aerial World. 8vo. 10s. 6d.
— Polar World. 8vo. 10s. 6d.
— Sea and its Living Wonders. 8vo. 10s. 6d.
— Subterranean World. 8vo. 10s. 6d.
— Tropical World. 8vo. 10s. 6d.
Haughton's Principles of Animal Mechanics. 8vo. 21s.
Heer's Primæval World of Switzerland. 2 vols. 8vo. 28s.
Helmholtz's Lectures on Scientific Subjects. 8vo. 12s. 6d.
Helmholtz on the Sensations of Tone, by Ellis. 8vo. 36s.
Hemsley's Handbook of Trees, Shrubs, & Herbaceous Plants. Medium 8vo. 12s.
Hullah's Lectures on the History of Modern Music. 8vo. 8s. 6d.
— Transition Period of Musical History. 8vo. 10s. 6d.
Keller's Lake Dwellings of Switzerland, by Lee. 2 vols. royal 8vo. 42s.
Kirby and Spence's Introduction to Entomology. Crown 8vo. 5s.
Lloyd's Treatise on Magnetism. 8vo. 10s. 6d.
— — on the Wave-Theory of Light. 8vo. 10s. 6d.
London's Encyclopædia of Plants. 8vo. 42s.
Lubbock on the Origin of Civilisation & Primitive Condition of Man. 8vo. 18s.
Nicols' Puzzle of Life. Crown 8vo. 3s. 6d.
Owen's Comparative Anatomy and Physiology of the Vertebrate Animals. 3 vols. 8vo. 73s. 6d.
Proctor's Light Science for Leisure Hours. 2 vols. crown 8vo. 7s. 6d. each.
Rivers's Rose Amateur's Guide. Fcp. 8vo. 4s. 6d.
Stanley's Familiar History of Birds. Fcp. 8vo. 3s. 6d.
Text-Books of Science, Mechanical and Physical.
    Abney's Photography, small 8vo. 3s. 6d.
    Anderson's Strength of Materials, 3s. 6d.
    Armstrong's Organic Chemistry, 3s. 6d.
    Barry's Railway Appliances, 3s. 6d.
    Bloxam's Metals, 3s. 6d.
    Goodeve's Elements of Mechanism, 3s. 6d.
    — Principles of Mechanics, 3s. 6d.
    Gore's Electro-Metallurgy, 6s.
    Griffin's Algebra and Trigonometry, 3s. 6d.

London, LONGMANS & CO.

Text-Books of Science—*continued.*

    Jenkin's Electricity and Magnetism, 3*s*. 6*d*.
    Maxwell's Theory of Heat, 3*s*. 6*d*.
    Merrifield's Technical Arithmetic and Mensuration, 3*s*. 6*d*.
    Miller's Inorganic Chemistry, 3*s*. 6*d*.
    Preece & Sivewright's Telegraphy, 3*s*. 6*d*.
    Shelley's Workshop Appliances, 3*s*. 6*d*.
    Thomé's Structural and Physiological Botany, 6*s*.
    Thorpe's Quantitative Chemical Analysis, 4*s*. 6*d*.
    Thorpe & Muir's Qualitative Analysis, 3*s*. 6*d*.
    Tilden's Chemical Philosophy, 3*s*. 6*d*.
    Unwin's Machine Design, 3*s*. 6*d*.
    Watson's Plane and Solid Geometry, 3*s*. 6*d*.

Tyndall on Sound. Crown 8vo. 10*s*. 6*d*.
— Contributions to Molecular Physics. 8vo. 16*s*.
— Fragments of Science. Crown 8vo.
— Heat a Mode of Motion. Crown 8vo.
— Lectures on Electrical Phenomena. Crown 8vo. 1*s*. sewed, 1*s*. 6*d*. cloth.
— Lectures on Light. Crown 8vo. 1*s*. sewed, 1*s*. 6*d*. cloth.
— Lectures on Light delivered in America. Crown 8vo. 7*s*. 6*d*.
— Lessons in Electricity. Crown 8vo. 2*s*. 6*d*.

Von Cotta on Rocks, by Lawrence. Post 8vo. 14*s*.
Woodward's Geology of England and Wales. Crown 8vo. 14*s*.
Wood's Bible Animals. With 112 Vignettes. 8vo. 14*s*.
— Homes Without Hands. 8vo. 14*s*.
— Insects Abroad. 8vo. 14*s*.
— Insects at Home. With 700 Illustrations. 8vo. 14*s*.
— Out of Doors, or Articles on Natural History. Crown 8vo. 7*s*. 6*d*.
— Strange Dwellings. With 60 Woodcuts. Crown 8vo. 7*s*. 6*d*.

CHEMISTRY & PHYSIOLOGY.

Auerbach's Anthracen, translated by W. Crookes, F.R.S. 8vo. 12*s*.
Buckton's Health in the House; Lectures on Elementary Physiology. Fcp. 8vo. 2*s*.
Crookes's Handbook of Dyeing and Calico Printing. 8vo. 42*s*.
— Select Methods in Chemical Analysis. Crown 8vo. 12*s*. 6*d*.
Kingzett's Animal Chemistry. 8vo.   [*In the press.*
— History, Products and Processes of the Alkali Trade. 8vo. 12*s*.
Miller's Elements of Chemistry, Theoretical and Practical. 3 vols. 8vo. Part I. Chemical Physics, 16*s*. Part II. Inorganic Chemistry, 24*s*. Part III. Organic Chemistry, New Edition in the press.
Watts's Dictionary of Chemistry. 7 vols. medium 8vo. £10. 16*s*. 6*d*.

THE FINE ARTS & ILLUSTRATED EDITIONS.

Doyle's Fairyland; Pictures from the Elf-World. Folio, 15*s*.
Jameson's Sacred and Legendary Art. 6 vols. square crown 8vo.
    Legends of the Madonna. 1 vol. 21*s*.
    — — — Monastic Orders. 1 vol. 21*s*.
    — — — Saints and Martyrs. 2 vols. 31*s*. 6*d*.
    — — — Saviour. Completed by Lady Eastlake. 2 vols. 42*s*.

London, LONGMANS & CO.

Longman's Three Cathedrals Dedicated to St. Paul. Square crown 8vo. 21s.
Macaulay's Lays of Ancient Rome. With 90 Illustrations. Fcp. 4to. 21s.
Macfarren's Lectures on Harmony. 8vo. 12s.
Miniature Edition of Macaulay's Lays of Ancient Rome. Imp. 16mo. 10s. 6d.
Moore's Irish Melodies. With 161 Plates by D. Maclise, R.A. Super-royal 8vo. 21s.
— Lalla Rookh. Tenniel's Edition. With 68 Illustrations. Fcp. 4to. 21s.
Redgrave's Dictionary of Artists of the English School. 8vo. 16s.

THE USEFUL ARTS, MANUFACTURES &c.

Bourne's Catechism of the Steam Engine. Fcp. 8vo. 6s.
— Handbook of the Steam Engine. Fcp. 8vo. 9s.
— Recent Improvements in the Steam Engine. Fcp. 8vo. 6s.
— Treatise on the Steam Engine. 4to. 42s.
Cresy's Encyclopædia of Civil Engineering. 8vo. 42s.
Culley's Handbook of Practical Telegraphy. 8vo. 16s.
Eastlake's Household Taste in Furniture, &c. Square crown 8vo. 14s.
Fairbairn's Useful Information for Engineers. 3 vols. crown 8vo. 31s. 6d.
— Applications of Cast and Wrought Iron. 8vo. 16s.
Gwilt's Encyclopædia of Architecture. 8vo. 52s. 6d.
Hobson's Amateur Mechanics Practical Handbook. Crown 8vo. 2s. 6d.
Hoskold's Engineer's Valuing Assistant. 8vo. 31s. 6d.
Kerl's Metallurgy, adapted by Crookes and Röhrig. 3 vols. 8vo. £4. 19s.
Loudon's Encyclopædia of Agriculture. 8vo. 21s.
— — — Gardening. 8vo. 21s.
Mitchell's Manual of Practical Assaying. 8vo. 31s. 6d.
Northcott's Lathes and Turning. 8vo. 18s.
Payen's Industrial Chemistry, translated from Stohmann and Engler's German Edition, by Dr. J. D. Barry. Edited by B. H. Paul, Ph.D. 8vo. 42s.
Stoney's Theory of Strains in Girders. Roy. 8vo. 36s.
Ure's Dictionary of Arts, Manufactures, & Mines. 4 vols. medium 8vo. £7. 7s.

RELIGIOUS & MORAL WORKS.

Arnold's (Rev. Dr. Thomas) Sermons. 6 vols. crown 8vo. 5s. each.
Bishop Jeremy Taylor's Entire Works. With Life by Bishop Heber. Edited by the Rev. C. P. Eden. 10 vols. 8vo. £5. 5s.
Boultbee's Commentary on the 39 Articles. Crown 8vo. 6s.
Browne's (Bishop) Exposition of the 39 Articles. 8vo. 16s.
Colenso on the Pentateuch and Book of Joshua. Crown 8vo. 6s.
Colenso's Lectures on the Pentateuch and the Moabite Stone. 8vo. 12s.
Conybeare & Howson's Life and Letters of St. Paul :—
    Library Edition, with all the Original Illustrations, Maps, Landscapes on Steel, Woodcuts, &c. 2 vols. 4to. 42s.
    Intermediate Edition, with a Selection of Maps, Plates, and Woodcuts. 2 vols. square crown 8vo. 21s.
    Student's Edition, revised and condensed, with 46 Illustrations and Maps. 1 vol. crown 8vo. 9s.
D'Aubigné's Reformation in Europe in the Time of Calvin. 8 vols. 8vo. £6. 12s.
Drummond's Jewish Messiah. 8vo. 15s.

London, LONGMANS & CO.

Ellicott's (Bishop) Commentary on St. Paul's Epistles. 8vo. Galatians, 8s. 6d. Ephesians, 8s. 6d. Pastoral Epistles, 10s. 6d. Philippians, Colossians, and Philemon, 10s. 6d. Thessalonians, 7s. 6d.
Ellicott's Lectures on the Life of our Lord. 8vo. 12s.
Ewald's History of Israel, translated by Carpenter. 5 vols. 8vo. 63s.
— Antiquities of Israel, translated by Solly. 8vo. 12s. 6d.
Goldziher's Mythology among the Hebrews. 8vo. 16s.
Jukes's Types of Genesis. Crown 8vo. 7s. 6d.
— Second Death and the Restitution of all Things. Crown 8vo. 3s. 6d.
Kalisch's Bible Studies. PART I. the Prophecies of Balaam. 8vo. 10s. 6d.
— — — PART II. the Book of Jonah. 8vo. 10s. 6d.
Keith's Evidence of the Truth of the Christian Religion derived from the Fulfilment of Prophecy. Square 8vo. 12s. 6d. Post 8vo. 6s.
Kuenen on the Prophets and Prophecy in Israel. 8vo. 21s.
Lyra Germanica. Hymns translated by Miss Winkworth. Fcp. 8vo. 5s.
Manning's Temporal Mission of the Holy Ghost. 8vo. 8s. 6d.
Martineau's Endeavours after the Christian Life. Crown 8vo. 7s. 6d.
— Hymns of Praise and Prayer. Crown 8vo. 4s. 6d. 32mo. 1s. 6d.
— Sermons; Hours of Thought on Sacred Things. Crown 8vo. 7s. 6d.
Mill's Three Essays on Religion. 8vo. 10s. 6d.
Monsell's Spiritual Songs for Sundays and Holidays. Fcp. 8vo. 5s. 18mo. 2s.
Müller's (Max) Lectures on the Science of Religion. Crown 8vo. 10s. 6d.
O'Conor's New Testament Commentaries. Crown 8vo. Epistle to the Romans, 3s. 6d. Epistle to the Hebrews, 4s. 6d. St. John's Gospel, 10s. 6d.
One Hundred Holy Songs, &c. Square fcp. 8vo. 2s. 6d.
Passing Thoughts on Religion. By Miss Sewell. Fcp. 8vo. 3s. 6d.
Sewell's (Miss) Preparation for the Holy Communion. 32mo. 3s.
Shipley's Ritual of the Altar. Imperial 8vo. 42s.
Supernatural Religion. 3 vols. 8vo. 38s.
Thoughts for the Age. By Miss Sewell. Fcp. 8vo. 3s. 6d.
Vaughan's Trident, Crescent, and Cross; the Religious History of India. 8vo. 9s. 6d.
Whately's Lessons on the Christian Evidences. 18mo. 6d.
White's Four Gospels in Greek, with Greek-English Lexicon. 32mo. 5s.

TRAVELS, VOYAGES &c.

Ball's Alpine Guide. 3 vols. post 8vo. with Maps and Illustrations:—I. Western Alps, 6s. 6d. II. Central Alps, 7s. 6d. III. Eastern Alps, 10s. 6d. Or in Ten Parts, 2s. 6d. each.
Ball on Alpine Travelling, and on the Geology of the Alps, 1s. Each of the Three Volumes of the *Alpine Guide* may be had with this Introduction prefixed, price 1s. extra.
Baker's Rifle and the Hound in Ceylon. Crown 8vo. 7s. 6d.
— Eight Years in Ceylon. Crown 8vo. 7s. 6d.
Brassey's Voyage in the Yacht 'Sunbeam.' 8vo. 21s.
Edwards's (A. B.) Thousand Miles up the Nile. Imperial 8vo. 42s.
Edwards's (M. B.) Year in Western France. Crown 8vo. 10s. 6d.
Evans's Through Bosnia and Herzegovina during the Insurrection. 8vo. 18s.
— Illyrian Letters. Post 8vo. 7s. 6d.

Grohman's Tyrol and the Tyrolese. Crown 8vo. 6s.
Hinchliff's Over the Sea and Far Away. Medium 8vo. 21s.
Indian Alps (The). By a Lady Pioneer. Imperial 8vo. 42s.
Lefroy's Discovery and Early Settlement of the Bermuda Islands. Vol. I. Royal 8vo. 30s.
Noble's Cape and South Africa. Fcp. 8vo. 3s. 6d.
Packe's Guide to the Pyrenees, for Mountaineers. Crown 8vo. 7s. 6d.
The Alpine Club Map of Switzerland. In four sheets. 42s.
Wood's Discoveries at Ephesus. Imperial 8vo. 63s.

WORKS OF FICTION.

Becker's Charicles; Private Life among the Ancient Greeks. Post 8vo. 7s. 6d.
— Gallus; Roman Scenes of the Time of Augustus. Post 8vo. 7s. 6d.
Cabinet Edition of Stories and Tales by Miss Sewell :—

Amy Herbert, 2s. 6d.
Cleve Hall, 2s. 6d.
The Earl's Daughter, 2s. 6d.
Experience of Life, 2s. 6d.
Gertrude, 2s. 6d.
Ivors, 2s. 6d.
Katharine Ashton, 2s. 6d.
Laneton Parsonage, 3s. 6d.
Margaret Percival, 3s. 6d.
Ursula, 3s. 6d.

Novels and Tales by the Right Hon. the Earl of Beaconsfield. Cabinet Edition, complete in Ten Volumes, crown 8vo. price £3.

Lothair, 6s.
Coningsby, 6s.
Sybil, 6s.
Tancred, 6s.
Venetia, 6s.
Henrietta Temple, 6s.
Contarini Fleming, 6s.
Alroy, Ixion, &c. 6s,
The Young Duke, &c. 6s.
Vivian Grey, 6s.

The Atelier du Lys; or, an Art Student in the Reign of Terror. By the Author of 'Mademoiselle Mori.' Crown 8vo. 6s.

The Modern Novelist's Library. Each Work in crown 8vo. A Single Volume, complete in itself, price 2s. boards, or 2s. 6d. cloth :—

By Lord Beaconsfield.
    Lothair.
    Coningsby.
    Sybil.
    Tancred.
    Venetia.
    Henrietta Temple.
    Contarini Fleming.
    Alroy, Ixion, &c.
    The Young Duke, &c.
    Vivian Grey.
By Anthony Trollope.
    Barchester Towers.
    The Warden.
By the Author of 'the Rose Garden.'
    Unawares.

By Major Whyte-Melville.
    Digby Grand.
    General Bounce.
    Kate Coventry.
    The Gladiators.
    Good for Nothing.
    Holmby House.
    The Interpreter.
    The Queen's Maries.
By the Author of 'the Atelier du Lys.'
    Mademoiselle Mori.
By Various Writers.
    Atherstone Priory.
    The Burgomaster's Family.
    Elsa and her Vulture.
    The Six Sisters of the Valley.

Lord Beaconsfield's Novels and Tales. 10 vols. cloth extra, gilt edges, 30s.
Whispers from Fairy Land. By the Right Hon. E. H. Knatchbull-Hugessen M.P. With Nine Illustrations. Crown 8vo. 3s. 6d.
Higgledy-Piggledy; or, Stories for Everybody and Everybody's Children. By the Right Hon. E. M. Knatchbull-Hugessen, M.P. With Nine Illustrations from Designs by R. Doyle. Crown 8vo. 3s. 6d.

London, LONGMANS & CO.

POETRY & THE DRAMA.

Bailey's Festus, a Poem. Crown 8vo. 12s. 6d.
Bowdler's Family Shakspeare. Medium 8vo. 14s. 6 vols. fcp. 8vo. 21s.
Conington's Æneid of Virgil, translated into English Verse. Crown 8vo. 9s.
Cayley's Iliad of Homer, Homometrically translated. 8vo. 12s. 6d.
Ingelow's Poems. First Series. Illustrated Edition. Fcp. 4to. 21s.
Macaulay's Lays of Ancient Rome, with Ivry and the Armada. 16mo. 3s. 6d.
Poems. By Jean Ingelow. 2 vols. fcp. 8vo. 10s.
    First Series. 'Divided,' 'The Star's Monument,' &c. 5s.
    Second Series. 'A Story of Doom,' 'Gladys and her Island,' &c. 5s.
Southey's Poetical Works. Medium 8vo. 14s.
Yonge's Horatii Opera, Library Edition. 8vo. 21s.

RURAL SPORTS, HORSE & CATTLE MANAGEMENT &c.

Blaine's Encyclopædia of Rural Sports. 8vo. 21s.
Dobson on the Ox, his Diseases and their Treatment. Crown 8vo. 7s. 6d.
Fitzwygram's Horses and Stables. 8vo. 10s. 6d.
Francis's Book on Angling, or Treatise on Fishing. Post 8vo. 15s.
Malet's Annals of the Road, and Nimrod's Essays on the Road. Medium 8vo. 21s.
Miles's Horse's Foot, and How to Keep it Sound. Imperial 8vo. 12s. 6d.
— Plain Treatise on Horse-Shoeing. Post 8vo. 2s. 6d.
— Stables and Stable-Fittings. Imperial 8vo. 15s.
— Remarks on Horses' Teeth. Post 8vo. 1s. 6d.
Nevile's Horses and Riding. Crown 8vo. 6s.
Reynardson's Down the Road. Medium 8vo. 21s.
Ronalds's Fly-Fisher's Entomology. 8vo. 14s.
Stonehenge's Dog in Health and Disease. Square crown 8vo. 7s. 6d.
— Greyhound. Square crown 8vo. 15s.
Youatt's Work on the Dog. 8vo. 12s. 6d.
— — — Horse. 8vo. 6s.
Wilcocks's Sea-Fisherman. Post 8vo. 12s. 6d.

WORKS OF UTILITY & GENERAL INFORMATION.

Acton's Modern Cookery for Private Families. Fcp. 8vo. 6s.
Black's Practical Treatise on Brewing. 8vo. 10s. 6d.
Bull on the Maternal Management of Children. Fcp. 8vo. 2s. 6d.
Bull's Hints to Mothers on the Management of their Health during the Pregnancy and in the Lying-in Room. Fcp. 8vo. 2s. 6d.
Campbell-Walker's Correct Card, or How to Play at Whist. 32mo. 2s. 6d.
Crump's English Manual of Banking. 8vo. 15s.
Cunningham's Conditions of Social Well-Being. 8vo. 10s. 6d.
Handbook of Gold and Silver, by an Indian Official. 8vo. 12s. 6d.
Longman's Chess Openings. Fcp. 8vo. 2s. 6d.

London, LONGMANS & CO.

Macleod's Economics for Beginners. Small crown 8vo. 2s. 6d.
— Theory and Practice of Banking. 2 vols. 8vo. 26s.
— Elements of Banking. Crown 8vo. 7s. 6d.
M'Culloch's Dictionary of Commerce and Commercial Navigation. 8vo. 63s.
Maunder's Biographical Treasury. Fcp. 8vo. 6s.
— Historical Treasury. Fcp. 8vo. 6s.
— Scientific and Literary Treasury. Fcp. 8vo. 6s.
— Treasury of Bible Knowledge. Edited by the Rev. J. Ayre, M.A. Fcp. 8vo. 6s.
— Treasury of Botany. Edited by J. Lindley, F.R.S. and T. Moore, F.L.S. Two Parts, fcp. 8vo. 12s.
— Treasury of Geography. Fcp. 8vo. 6s.
— Treasury of Knowledge and Library of Reference. Fcp. 8vo. 6s.
— Treasury of Natural History. Fcp. 8vo. 6s.
Pewtner's Comprehensive Specifier; Building-Artificers' Work. Conditions and Agreements. Crown 8vo. 6s.
Pierce's Three Hundred Chess Problems and Studies. Fcp. 8vo. 7s. 6d.
Pole's Theory of the Modern Scientific Game of Whist. Fcp. 8vo. 2s. 6d.
The Cabinet Lawyer; a Popular Digest of the Laws of England. Fcp. 8vo. 9s.
Willich's Popular Tables for ascertaining the Value of Property. Post 8vo. 10s.
Wilson's Resources of Modern Countries 2 vols. 8vo. 24s.

MUSICAL WORKS BY JOHN HULLAH, LL.D.

Chromatic Scale, with the Inflected Syllables, on Large Sheet. 1s. 6d.
Card of Chromatic Scale. 1d.
Exercises for the Cultivation of the Voice. For Soprano or Tenor, 2s. 6d.
Grammar of Musical Harmony. Royal 8vo. 2 Parts, each 1s. 6d.
Exercises to Grammar of Musical Harmony. 1s.
Grammar of Counterpoint. Part I. super-royal 8vo. 2s. 6d.
Hullah's Manual of Singing. Parts I. & II. 2s. 6d.; or together, 5s.
Exercises and Figures contained in Parts I. and II. of the Manual. Books I. & II. each 8d.
Large Sheets, containing the Figures in Part I. of the Manual. Nos. 1 to 8 in a Parcel. 6s.
Large Sheets, containing the Exercises in Part I. of the Manual. Nos. 9 to 40, in Four Parcels of Eight Nos. each, per Parcel. 6s.
Large Sheets, the Figures in Part II. Nos. 41 to 52 in a Parcel, 9s.
Hymns for the Young, set to Music. Royal 8vo. 8d.
Infant School Songs. 6d.
Notation, the Musical Alphabet. Crown 8vo. 6d.
Old English Songs for Schools, Harmonised. 6d.
Rudiments of Musical Grammar. Royal 8vo. 8s.
School Songs for 2 and 3 Voices. 2 Books, 8vo. each 6d.
Time and Tune in the Elementary School. Crown 8vo. 2s. 6d.
Exercises and Figures in the same. Crown 8vo. 1s. or 2 Parts, 6d each.
Helmore's Catechism of Music, based by permission on Dr. Hullah's Method. Crown 8vo. 1s. sewed in paper; or 1s. 2d. sewed in canvas.

London, LONGMANS & CO.

www.ingramcontent.com/pod-product-compliance
Lightning Source LLC
Chambersburg PA
CBHW021943240426
43668CB00037B/491